IS CAPITALISM SUSTAINABLE?

DEMOCRACY AND ECOLOGY
A Guilford Series
Published in Conjunction with
the Center for Political Ecology

James O'Connor
Series Editor

IS CAPITALISM SUSTAINABLE?
Political Economy and the Politics of Ecology
Martin O'Connor, *Editor*

Forthcoming Topics and Authors/Editors
Ted Benton, **Marxism and Ecology**
David Macauley, **Philosophers of Nature**
Daniel Faber, **Radical Environmental Movements Today**
James O'Connor, **Political Economy of Nature**
Enrique Leff, **Green Production**

Is Capitalism Sustainable?
Political Economy and the
Politics of Ecology

Edited by
MARTIN O'CONNOR

Introduction to the Series by
James O'Connor

THE GUILFORD PRESS / New York   London

© 1994 The Guilford Press
A Division of Guilford Publications, Inc.
72 Spring Street, New York, NY 10012

Printed in the United States of America

This book is printed on acid-free paper.

Last digit is print number: 9   8   7   6   5   4   3   2

Library of Congress Cataloging-in-Publication Data

Is capitalism sustainable?: political economy and the politics of
    ecology / edited by Martin O'Connor.
        p.   cm. — (Democracy and ecology)
    Includes bibliographical references and index.
    ISBN 0-89862-127-5 (hardcover). — ISBN 0-89862-594-7 (pbk.)
    1. Economic development—Environmental aspects. 2. Sustainable
development. 3. Environmental policy. I. O'Connor, Martin.
II. Series.
HD75.6.I82   1994
363.7—dc20   94-11689
                                                                CIP

Contributors

JUAN MARTÍNEZ ALIER, Department of Economics and Economic History, Universidad Autónoma, Barcelona, Spain

ELMAR ALTVATER, Free University, Berlin, Germany

FRANK BECKENBACH, Department of Social Sciences, University of Osnabrueck, Osnabrueck, Germany

JEAN-PAUL DELÉAGE, Editor, *Écologie Politique*; Department of Environment, Université de Paris, Paris, France

ALEX DEMIROVIĆ, Department of Social Sciences, Johann Wolfgang Goethe-Universität and the Institut für Sozialforschung, Frankfurt, Germany

JOHN S. DRYZEK, Department of Political Science, University of Oregon, Eugene, Oregon

MARGARET FITZSIMMONS, Graduate School of Architecture and Urban Planning, University of California, Los Angeles, California

MICHAEL GISMONDI, Department of Sociology and Global Studies, Athabasca University, Athabasca, Alberta, Canada

JOSEPH GLASER, Los Angeles, California

COLIN HAY, Department of Sociology, Lancaster University, Lancaster, United Kingdom

ROBERTO MONTE MOR, CEDEPLAR, Federal University of Minas Gerais, Belo Horizonte, Brazil

JAMES O'CONNOR, Editor-in-Chief, *Capitalism, Nature, Socialism*; Department of Sociology and Economics (Emeritus), University of California, Santa Cruz, California

MARTIN O'CONNOR, Department of Economics, University of Auckland, Auckland, New Zealand

STEPHANIE PINCETL, Department of Geography, University of California, Berkeley, California

SUDHIR CHELLA RAJAN, Environmental Science and Engineering, University of California, Los Angeles, California

MARY RICHARDSON, Department of Philosophy and Humanities, Athabasca University, Athabasca, Alberta, Canada

ARIEL SALLEH, Ecofeminist activitist, Glebe, New South Wales, Australia; Convener of the Women's Environmental Education Centre, Sydney, Australia; Occasional Visiting Scholar, Environmental Conservation Education Program, New York University, New York, New York

Introduction to the Democracy and Ecology Series

This book series titled "Democracy and Ecology" is a contribution to the debates on the future of the global environment, the "free market economy," and the prospect of radical green and democratic movements in the world today. While some call the post-cold war period the "end of history," others sense that we may be living at its beginning. Among these are the editors of the volumes of this series, and most of the authors whose works are presented therein, who believe that our species is learning to regulate the relationship that we have with the rest of nature in ways that defend ecological values and sensibilities and that right the exploitation and injustice that disfigure the present world order. Most would concur that ecological and social wrongs are closely interconnected and that solutions to both presuppose radical forms of political democracy and new values that depart from those underpinning the present capitalist system that supposedly has closed the book on ideology and leftist politics. All are asking hard questions about what went wrong with the worlds that global capitalism and state socialism made and the kind of worlds that might be rebuilt from the wreckage of ecologically and socially bankrupt ways of life. The Democracy and Ecology series rehearses these questions, poses new ones, and tries to answer them, if only tentatively and provisionally, because the stakes are so high, and since "time-honored slogans and time-worn formulae" have become part of the problem.

What are "ecology" and "democracy"? How do these words work in dominant and critical discourses on nature and politics? Some define *ecology* as "ecological science," others as "ecological movement." For some, "ecology" still means the "balance of nature," for others, that "everything is connected to everything else," for still others, "self-organization and self-creation" and "chaotic evolution." Corporate planners, seeking to tap into nature's fecundity and human creativity, dream of remaking both nature and human labor power into "natural" and "human capital," respectively. International civil servants imagine an "ecotocracy," a bureaucracy of ecological scientists and technicians, to manage the endangered world's commons; urban planners make blueprints for green cities; deep ecologists fantasize systems of material production based on "bioregionalism." "Ecology" is used in defense of, and also against, the ways of life of indigenous peoples and peasant cultures; as a catchall term denoting environmental preservation and restoration; as a descriptive for New Age naturalistic spirituality; and as a term emblematic of green movements seeking social justice with as much fervor as environmental sustainability.

Democracy too is associated with a variety of economic and political agendas. In classical Athens originally meaning "popular rule" or "revolution," in liberal democracies the term denotes "democratic procedures for electing representatives to law-making bodies." "Ecocrats" reject this definition on the grounds that party competition, lobbying by special interests, and other features of liberal democracy work too slowly and chaotically to meet the challenge of the global environmental crisis. Most greens agree that an "ecological society" cannot be constructed within liberal democratic political forms, but vehemently oppose top-down "ecocratic" planning in favor of a devolution of power and regional or local environmental decision making. Between ecocrats and greens stand the "moderates," who urge national and international bureaucracies responsible for environmental regulation to be less secretive and more accessible and accountable to the "public."

The range of political views among "left greens" is equally wide. Barry Commoner, one of the founding fathers of left ecology in the United States, calls for "social governance of production," while Murray Bookchin, the other founder, supports "libertarian municipalism" and "direct democracy." In poor countries, left green movements in the countryside fight to "delink" villages from the division of global labor and in the cities to empower urban neighborhoods. Most, if not all, left greens (like plain greens) tend to define democracy in terms of local power or communalism and to support "identity politics" or the "politics of place." Some "red greens," however, are thinking about how to supersede or "sublate" exclusive and inclusive cultural identities; "local" and "central" concepts of place; spontaneity and planning; the divisions of industrial and social labor; "farmers" science and university science; site-specific and abstract knowledge; and even atomistic and holistic elements within ecological science itself. The debates within the women's movement, especially left ecological feminism, which have problematized "democracy" to include the organization of the labor of reproduction, also hold promise for resolving key issues of democracy and ecology—the problem of scale; the meaning of "site specificity"; the forms of social organization that best fit the household, town, or village, region, nation, and world as a whole. Some have suggested that ecology (however defined) knows no national boundaries and hence tends to subvert and revolutionize dominant political and economic institutions (e.g., the nation-state and transnational corporations and banks) and to create the need for diverse and multilevel democratic international, continental, bioregional, and local political and economic bodies and practices. A radical extension, and redefinition, of democracy might (others claim) "naturally" embrace a biocentric ethic. Nature philosophy is coming down to earth.

All radical greens, deep ecologists, socialist biocentrics, green anarchists and socialists, and ecological feminists subscribe to the view that global capitalism is not sustainable, however "sustainability" is defined. Wide agreement exists that the world economy is producing greater poles of wealth and poverty and that the weight of competition, national and regional rivalries, and the accumulation imperative, on the one hand, and the absence of any rational, overall planning, on the other hand, mitigate against "green capitalism." Most concur that ecological and economic crises feed on one another and that both are linked to the growing concentration of political power and privilege.

Some of us, however, think that seemingly all-powerful, reified world of global capital is creating more ecological, economic, social, and political problems than the world's ruling and political classes will be able to solve. We share the feeling that a general crisis is over the horizon. This turning point or divide (the argument goes) will create opportunities, as well as dangers, if the social movements that are addressing the problem of humankind's relation to nature and to itself are able to develop politically strategic, as well as tactical, alliances, locally, regionally, and internationally. Our slogan might be "Think and act locally and globally." Ultimately, we think that there may be an underlying "fit" between ecology, defined as the self-organization and self-creation of life, and democracy, defined as "self-governance." Hence, our sense that history may be beginning, not ending.

JAMES O'CONNOR
Series Editor

Contents

IS CAPITALISM SUSTAINABLE?

Gentle breath of yours my sails
Must fill, or else my project fails,
Which was to please. For now I want
Spirits t'enforce, art to enchant;
And my ending is despair,
Unless I be relieved by prayer;
Which pierces so, that it assaults
Mercy it self, and frees all faults.
   As you from crimes would pardon'd be,
   Let your indulgence set me free.

—PROSPERO, in the Epilogue to
   The Tempest, by William Shakespeare

Introduction: Liberate, Accumulate—and Bust?

Martin O'Connor

1. Ecological Crisis in Market Society

For capitalism, sustainability is taken to mean sustained growth; technology is the vehicle of an irreversible ascension; and accumulation is asserted to be a destiny for all individuals and nations who work hard at it. The hallmark of bourgeois society—as theorized by Mandeville, Locke, and many others since—is the legitimation of "self-interest." Private vice is public virtue; individual avarice and profit seeking are the engines of social progress; and the good and just society is achieved by nothing other than the establishment of good order among a multitude of self-interested persons producing, accumulating, and exchanging in the marketplace.

With the collapse of the Soviet hegemony and of Eastern bloc communism, the ascendancy of this image of the good society would appear unassailable. Yet, in real terms, world capitalism is assailed by wrenching contradictions. The title question—"Is capitalism sustainable?"—encapsulates in an ironical way the sorts of questions that are on everybody's lips since the Rio de Janeiro conference in 1992. First, given what we know about capitalist relations and forces of production, and about the failures of capitalism up until now to ensure regeneration of the social and ecological conditions needed for accumulation, is an *ecologically sustainable* capitalism possible? Second, and more crucial, can one plausibly suggest that capitalism might be reformed to respect the integrity of the social and ecological domains hitherto subject to savage exploitation?

The examination of capitalist economic processes and of liberal democratic institutions as they exist leads for both questions to the answer, "Probably not." This leaves the implication that, although capitalism on a world scale may be viable for some time yet to come, the prices to be paid in terms of both human conflicts and lost human, cultural, and ecological riches will be very great. The contributions to this volume confront that prospect. They are written across the knife edges of hope and despair, utopianism, urgency for change, and compromises of practical politics. The hope nourishes the commitment to critical analysis, and sometimes the analysis tempers the hope.

The horrible reality of increasing numbers in absolute poverty worldwide is easily put down to unchecked population growth and the indigence of the poor. Yet, Third World populations now in disarray were not "originally" in some state of "primitive" ignorance and destitution. Most parts of what is now the rest of the world relative to the West have had, in the recent or more distant past, a variety of high civilizations. One is left to wonder what the legacy of European colonization (including the post-1945 "decolonization and development" phase) has been in making these peoples poor. Worldwide poverty is clearly not a question of as-yet-inadequate development of "productive forces" on a world scale. More than enough food is being produced on this planet to feed abundantly every body and soul. Far more "scarce resources" are squandered every day on military adventurism and arms stockpiling than are furnished in any sorts of "development assistance;" and the destruction of surplus foodstuffs and other commodities simply because they are unprofitable is a normal part of Western, capitalist business practice.

Not all the ills and imbalances of the world economy can be attributed to capitalism. Yet, for most people, the free market and modernity have so far produced little more than a phantom of freedom. The unequal distribution of wealth and power is a hallmark of liberal capitalism. As August Bebel once formulated it, "Economic liberalism is the free fox in the free chickencoop."[1] Serge Latouche comments that

> in spite of all its reforms and reinterpretations capitalism remains
> tainted by the accusation of exploitation. For the worker anyway,
> it is the incarnation neither of economic democracy nor of emanci-
> pation.[2]

Keeping a fair face on capitalism thus requires a continual patch-up job. Throughout the history of the market-based economic order, there has been a continual expenditure of energy (both physical and metaphorical) to legitimate "free enterprise" and to mitigate some of the more obvious contradictions and inequalities. Capitalist responses to environmental crisis can be seen in this light. Problems such as the despoliation of countrysides and the ill health of workers breathing industrial dust and poisons are nothing new. In the 19th century they had been remarked by Marx as a corollary of the competitive drive for profits; but Marx himself had contributed to the optimistic view that these problems were resolvable *technologically*, if only the *social* relations of production were appropriately reformed. The fear of natural resource depletion is as old as the idea of an expanding economy based on drawdown of God-given resources. The 19th-century thermodynamicists, extrapolating from the steam engine, developed "a conception of society and men as energy transforming engines"; and correspondingly they saw nature as "a reservoir of energy that is always threatened with exhaustion."[3] However, the discovery of new energy sources along with new techniques for harnessing them (e.g., from coal to oil and gas, to uranium nuclear fission, to fast-breeders and fusion) led many commentators of progress to reject the specter of resource depletion.

In short, it had been convenient—at both metaphysical and practical levels—to set ecological concerns to one side, with the proposition that further progress and accumulation will bring pollution-free technologies, new remedies for environmental damage, new resource discoveries, and improved workplace and leisure comforts to all. But today a growing consensus suggests that the spiraling problems of social and environmental costs will not go away.

The 1960s saw a growing agitation regarding the poisoning of urban and rural habitats occasioned by economic growth and consumerism; this habitat concern was overtaken in the 1970s by the "energy crisis" and its attendant preoccupation with natural resource depletion; and the 1990s promise to be the decade of the "global ecology" issues. David Ricardo at the beginning of the 19th century had been able to write of the "indestructible" powers of the land, and to aver that: "the brewer, the distiller, the dyer, make incessant use of their air and water for the production of their commodities; but as the supply is boundless, they bear no price."[4] By the 1980s, it was not longer remotely plausible to treat the raw materials and "services"—source, site, scenery, and sink—furnished by nature as indestructible and/ or nonscarce.

So, contemporary environmental problems represent not only a major economic crisis of supply, but also a new crisis of legitimacy for the market system. Popularly, there is less and less confidence in the phantasms of unbridled affluence that constitute the American Dream for the North and the Development Dream in the South.[5] Pamela Zoline, in a short story titled "The Heat Death of the Universe,"[6] which chronicles a day in the demise of a suburban housewife, captures with bleak irony this sense of hollowness and doubt. Sarah Boyle is described as "a vivacious and intelligent young wife and mother, educated at a fine Eastern college, proud of her growing family which keeps her busy and happy round the house." At morning breakfast, the author reports,

> With some reluctance Sarah Boyle dishes out Sugar Frosted Flakes to her children, already hearing the decay set in upon the little milk-white teeth, the bony whine of the dentist's drill. . . . One bowl per child. . . .
>
> The box blats promises: Energy, Nature's Own Goodness, an endless pubescence. On its back is a mask of William Shakespeare to be cut out, folded, worn by thousands of tiny Shakespeares in Kansas City, Detroit, Tucson, San Diego, Tampa. . . . Two or more of the children lay claim to the mask, but Sarah puts off that Solomon's decision until such time as the box is empty.
>
> A notice in orange flourishes states that a Surprise Gift is to be found somewhere in the package, nestled amongst the golden flakes. So far it has not been unearthed, and the children request more cereal than they wish to eat, great yellow heaps of it, to hurry the discovery. Even so, at the end of the meal, some layers of flakes remain in the box, and the Gift must still be among them.
>
> There is even a Special Offer of a secret membership code and magic ring; these to be obtained by sending in the box top with 50¢.
>
> Three offers on one cereal box. To Sarah Boyle this seems to be oversell. Perhaps something is terribly wrong with the cereal and it must be sold quickly, got off the shelves before the news breaks. Perhaps it causes a special cruel cancer in little children. As Sarah Boyle collects the bowls printed with bunnies and baseball statistics, still slopping half full of milk and wilted flakes, she imagines in her mind's eye the headlines, "Nation's Small Fry Stricken, Fate's Finger Sugar-Coated, Lethal Sweetness Socks Tots."

2. The "Contradictions" of Capitalist Accumulation

The purpose of this volume is to tease out some of the facets of capitalism's ecological contradiction, and to discuss critically the politics of ecology under capitalism. This means looking at the ways that capitalism and liberal politics themselves are responding to this legitimation crisis, and also at the grounds of social resistance to capitalist predations. The "contradictions" have both material and political/institutional dimensions. The essays collected here permit no single, hard-and-fast ordering; but the arrangement of the book reflects some structure within a multiplicity of angles of approach.

Contributions making up the first half of the book are primarily devoted to appraisal of the contradictions of capitalist accumulation in their ecological and economic dimensions. The contributors take up different theoretical aspects of the relationship between capitalism and nature—or, more exactly, of people with each other and with nonhuman nature under capitalism. The questions to be answered are: Is capitalist production, distribution, exchange, consumption, and accumulation consistent with ecological sustainability? And supposing it is, or might be made to be, what social and economic interests will be provided for, and what will be forcibly suppressed? Market society proclaims itself as offering an ever-expanding range of choices to humankind through its fostering of output growth, innovation, and the free (competitive) play of ideas and initiatives. Pretendedly, there is a harmonization of interests, with market forces establishing the "best use" of resources, and thus an "optimal balance" between competing needs. But the realities are very different. Borrowing Maria Mies's term and generalizing its application, there is a *superexploitation* both of the wealth of the earth and of human societies.[7] In ecological terms, there is an appropriation of the earth's fecundity as a "natural resource" in the service of accumulation, and a running down of this resource. In social terms, there is an appropriation of human nature, a domination exercised worldwide by capital (increasingly transnational: *offshore*) over humans qua labor and reproductive power. This means hegemony over industrial labor power, but also over domestic labor, over peasant (subsistence) labor in non-commodified domains, over communal work of social repair and regeneration, and over women's labors of human regeneration. In both respects—human and nonhuman—the domination does not entail mere appropriation of a surplus; rather, it is a destructive process whose result is the dereliction of human societies and ecosystems alike.

This process of spiraling depredation can be analyzed from a number of angles. One is in terms of the biophysical dimensions of economic activity, emphasizing the material and energy (inter)dependency of all economic production and consumption activity with its physical environment. This is the perspective taken up by Juan Martínez Alier, Jean-Paul Deléage, myself, and Elmar Altvater in the four essays that open the volume.

Juan Martínez Alier, in "Ecological Economics and Ecosocialism," gives a historical overview of the concerns of ecological or bioeconomics from the 19th century to the present. Reviewing scientific, ideological, and social issues of debate, he brings out three interrelated themes of bioeconomics: environmental scarcity, entropic irreversibility, and the problem of economic justice through time. Arguments about the roles or adequacy of physical–ecological science and economic analysis in defining policy action have behind them the urgent political and material problems of social, cultural, and economic domination—injustices in terms of access to the amenities and resources of our natural world, and in bearing risks and burdens of industrial fallout in years to come.

Jean-Paul Deléage, in his "Eco-Marxist Critique of Political Economy," makes the same connections in a slightly different way. Thermodynamic science in the 19th century gave an explanation, in terms of laws of nature, for phenomena that everyone understood intuitively: that coal, once burnt, cannot be restored for reuse, and that economic activity necessarily involves production of physical wastes. This is entropic irreversibility, as defined by the Second Law of Thermodynamics. Why is it a critical consideration in ecological economics? If natural resources were abundant, and sites for waste disposal were available without limit, entropic irreversibility would condition, but not threaten, expanding economic activity. As Deléage says, capitalism has always, in its *representation* of the accumulation process, treated nature as unlimited. But this way of representing things is becoming untenable given the rates of resource consumption and the changes in the terrestrial environment caused by extraction and waste-disposal activities. Environmental scarcity is here, and to stay.

These material "limits to growth" are one half of capitalism's ecological contradiction. The other half, as Deléage and Alier both intimate and as I discuss at length in "Codependency and Indeterminacy: A Critique of the Theory of Production," is a crisis of control. Industrialism has been premised on putting nature to productive use, that is, on an instrumental logic that presumes nature is harnessed (by labor and technology) to a desired end. But, as ecological and thermodynamic sciences are now teaching us, we *live with/in nature* rather than on (or off) her. Human economy and habitat is a relation of intimacy and mutual compatibility, which can be ruptured by ill-considered extractions, intrusions, or waste disposals. Technological advances sharpen our capabilities for calculated intervention in human and nonhuman systems—but they also heighten the stakes of the uncontrolled and incalculable dimensions of induced change.

If a basic principle of capitalist accumulation is controlled exploitation of nature (and of human labors), and this premise is false, then industrial "development" is now orienting the world economy toward the destruction of nature on a planetary scale. This provides a point of departure for Elmar Altvater in his analysis of "Ecological and Economic Modalities of Time and Space." Economic production is, in physical terms, transformation happening in space and time. The capitalist enterprise, however, is not concerned with material transformations as such. Entrepreneurs and executives are largely indifferent to the physical dimensions of such transformations; these matter only in terms of the requirement that their enterprise remains viable—so long as a capital surplus is achieved that permits them to set in motion further cycles of the transformation process that nets them a surplus.[8] It is possible to see straightaway the mismatch between this illusion of a self-feeding value-accumulation, and the material realities of this process—namely, the *equality* of inputs and outputs in brute energy and material terms, and the *qualitative irreversible changes* in entropy and ecological terms. Social and ecological processes unfold historically through the dimensions of space and time. In this unfolding, numerous time scales and spatial scales are relevant—from the glacial advances and retreats, to the lunar cycles and the motions of the planets around the sun, to the time taken to prepare and cook a meal. This is historical time, which contrasts sharply with the monetary reductionism in-built in capitalism, by which equivalents-in-value are defined in market exchanges. Altvater recalls the adage of the German Greens: "When the last tree has been cut, one will realize that one cannot eat money."

The contemporary critique of political economy must, therefore, turn from Marxism's traditional focus on the mechanisms of exploitation of labor power within growing industrial economies, to analyze, *in addition*, the social mechanisms of destruction and exploitation of nature. Essays by Frank Beckenbach, Ariel Salleh, myself again, and James O'Connor are efforts at this task. Here, the attempt is made to locate the industrial exploitation of wage labor within a larger context that also includes capitalist exploitation of domains "external" to the capitalist workplace but on which accumulation depends. These are, speaking abstractly, the domains of (1) biophysical nature (the environment, or the planet considered as source of raw materials, sink, and habitat); (2) human nature not only as a source of industrial labor power in wage work, but also as household labors, human reproduction, and noncommodity subsistence production; and (3) social infrastructures including transportation, delivery of health and educational services, law and order, and so on.

These domains may be referred to generically as capital's *conditions of production*[9]; *thus what is sought in the shift from industrial to ecological Marxism* is to develop a critique of the mechanisms of capital's degradation of these conditions of production. One can think of the domination exercised by capital over labor and nature, over work spaces and habitats, in two complementary senses—as predations, and as the imposition of unwanted burdens. First, there is "exploitation" in the classical sense of capital's obtaining its needed raw materials and services on a nonreciprocal basis: extraction of "free gifts of nature," and appropriation of the products of wage labor and of domestic and subsistence labors. Second, and just as real, is the exploitation of people as reluctant *recipients* of capitalism's debris, such as degraded working conditions, poisons, and fallout. In both instances there is reduction of people (and of nonhuman nature) to the status of subservient instruments, as mere factory floor, inputs, and cogs of the accumulation machine.

Frank Beckenbach, in his essay, "Social Costs in Modern Capitalism," develops his analysis by looking at manifestations of the "ecological contradiction" at the level of capitalist firms' operating imperatives. He gives a detailed discussion of the ways modern mixed economies are organized to achieve a systematic *cost shifting* from private capitalist enterprise onto the state and taxpayers, workers (in terms of bad working conditions, commuting costs, etc.), the public at large, future generations, and nonhuman nature. Competitive enterprises will seek to lower input costs and to offload costs onto other parties (government, communities, future generations, etc.) when profits and survival are under threat. He concludes that unless and until capitalist enterprises are obligated to take full account of the burdens they load onto communities and ecosystems—a responsibility that clearly conflicts with the profit motive—the degradation of working and living conditions will continue. At the same time, this cost-shifting will be resisted by the individuals and communities affected, leading to politicization of the issues. As James O'Connor has elsewhere formulated it:

> The combination of crisis-stricken capitals externalizing more costs, the reckless use of technology and nature for value realization in the sphere of circulation, and the like, must sooner or later lead to a "rebellion of nature," i.e., powerful social movements demanding an end to ecological exploitation. Especially in today's crisis, whatever its source, capital attempts to reduce production and circulation time, which typically has the effect of making environmental practices, health and safety practices, etc. worse. Hence capital restructuring may deepen not resolve ecological problems.[10]

Now, what appears, within capitalist economy, as degraded production conditions and difficulties of supply posing a threat to profitability and growth, is also, materially, the means and substance of life—human and nonhuman life. As Giovanna Ricoveri says, in an abstract sense the origin of the present ecological crisis lies in "the mortal conflict between capital and nature[11]; but

> people are also part of nature, and the exploitation of nature is therefore also the exploitation of some people by other people. Ecological degradation is also the degradation of human relationships.

This is a theme Ariel Salleh addresses directly, in "Nature, Woman, Labor, Capital: living the deepest contradiction." Her focus is the gendered division of labor and therefore the distinct perspectives that women bring to bear on the expansion of capitalist relations of production. Under capitalism considered as a modern form of patriarchal relations, women find themselves defined variously as "natural resource" or infrastructural "conditions of production." In the capitalist–patriarchal tradition, women's reproductive power locates them within "nature"; while their unpaid labor—subsistence chores, caring, household maintenance, and manual work in making goods as farmers or weavers—simply serves to "bridge" men (markets, militarism, management) and nature. While capitalist "development" has brought obvious material benefits to some, the majority of women living the *nature–woman–labor nexus* can expect only a worsening of environmental conditions and of material and psychological stresses.

The expansion worldwide of the "market economy" and its intrusion into hitherto autonomous domains—especially subsistence agriculture, household activity, and women's social labors—annexes women/nature to capital with little economic return. In the case of Western women, this is carried through by such mechanisms as medicalization of birth and childcare, "convenience" commodities, and so-called labor-saving gadgetry in the home.[12] In Third World societies, the annexation removes the basis for autonomous communal subsistence—by technocratic "revolutions," forced resettlement, the extraversion of production toward commodity markets, and expropriation tactics such as gene patenting by multinational corporations.

This mode of ecofeminist analysis identifying the *nature–woman–labor nexus* as a fundamental contradiction lends support to the ecological and thermodynamic view of capitalism as a predatory process. At the same time, it insists that the exploitation or nature has to be understood as a *social relation* with divisions of labor effected along lines of gender, ethnicity, and race as well as industrial class. By emphasizing the political import of the marginalized voice—grounded dialectically in "need" and everyday experiences of denial—ecofeminism makes visible an existing basis for resistance to capitalism and expresses an alternative social form.

My own essay, "On the Misadventures of Capitalist Nature," looks, at a fairly high level of abstraction, at the ideological mechanisms of silencing and dispossession that grease the way of capital's worldwide expansion. Capitalism is a *colonizing* force in a double way. The penetration of capital around the globe involves not just an invasion, with associated plunder and despoliation, but a semiotic conquest, a sort of outgrowth that, as in a tumor, progressively envelops its surroundings. In the rhetoric of "greened growth" and "sustainable development," we can, I suggest, observe a sinister double play around the categories of *nature/capital*, for the better legitimation of capitalist accumulation and relations of production. The first play is the traditional one of appropriation of territories—defining them as *nature* and as *nobody's property* before making them capital's own for purposes of resource extraction, rent seeking, construction projects, or waste disposal. The second is the entrapment of communities within the capitalist logic through the process of defining valuable sites or sources of services as "capitals," henceforth to be managed according to the imperatives of the (global) marketplace. People are shotgunned or seduced into conceiving of themselves as proprietors (or stewards) of themselves and their habitats *as capital* (human capital, ecological and genetic capital, tribal community assets, etc.) which they may choose either to conserve or to proffer in the marketplace. This is the semiotic conquest of the "natural" domain as a type of capital. The planet itself becomes modeled as "our" collective stock of capital having to be sustainably managed. Of course, real predation continues to take place; but now this takes the form of a reciprocal (but uneven) *cannibalism of capitals* where the "strong" eat the weak.

Slavery and the European colonial doctrine of *terra nullius* used in the past to dispossess peoples of their habitats and lives throughout Africa, the Americas, and Oceania show how far such frauds and monologues can go. The capitalization game is no less ruthless and effective. A striking example of dispossession is the patenting of seed and plant varieties derived from stocks used by Third World "traditional" societies.[13] A currently fashionable argument is that whoever "develops" a new strain can claim patent rights, and thereafter force payment from anyone making use of that seed strain. This involves drawing an imaginary boundary between a putative "natural" domain defined as open access, and a domain of "humanly produced" property that is privatized. The practical effect is to *dispossess* the farming communities who, over decades and centuries, have furnished and maintained as common property the genetic stocks that are now defined as constituting the "natural" raw materials for commercial breeding and hybridization.[14] Farmers using a patented seed would, under the new regimes being instituted, have to pay for the right to save and replant the seed stock each season, or for the right to use the patented seed in their own breeding efforts. On the other hand, the commercial patent holders pay nothing for their own continuing use and benefit from the "natural" germplasms they have taken and recombined. The elaboration of tradable property rights in such contexts is, Vandana Shiva suggests,

> part of the continuing process of "primitive accumulation," of a violent dispossession of autonomous producers of their means and conditions of production. And it is robbery created by science protected by law.[15]

Transnational capital thereby assures and legitimates its privileged access to needed raw materials and services, in the name of investments for the good of humanity. But the nature/property demarcations, and the patents thus claimed, are designed to protect and enhance the value of commercial investments; the subsistence communities become casualties of commercial war!

So, as James O'Connor asks, "Is Sustainable Capitalism Possible?" If by "sustaining" we simply mean "uphold the course" of capitalist accumulation and relations of production globally, then *in the long run*, "Probably not." And if by "sustaining" we mean to "provide the necessities of life" for the peoples of the world, and respect for those whose ways of life are presently being subverted by the wage and commodity forms, the answer seems flatly, "No." On the other hand, he suggests, one can hope for political resistance and action toward some kind of *ecological socialism*—the self-conscious development of a common or public sphere, a political space, a kind of dual power, in which minority, labor, women, urban, and environmental organizations can work economically and politically.

This brings us directly to questions of power and political process, particularly to the themes of *knowledge, forms of cooperation, democratization*, and *the state*. All the contributions in the book are concerned, one way or another, with questions of the experiential grounding of knowledge, the translation of knowledge into political action, and the construction of cooperative utopias. The essays making up the second half focus particularly on the political institutions of the liberal democracies, and their potential (and limits) as vehicles for effective resolution of capitalism's ecological contradictions. The rulers or governments of each state, as "sovereign powers" of defined territories, intermediate the contests over access to resources and environments (when they do not take direct "development" roles themselves). So, to what extent does it make sense to call upon the state in the struggle to make capitalism change its exploitative spots? This question does not have a ready answer. First, the nation state and the Western form of liberal (parliamentary) democracy are colonial legacies in themselves, and are not disjoint from that other Western avatar of liberty and (consumer) sovereignty: the market. Does this make them irremediably complicit with capital? Second, even supposing that the (liberal) states might be made responsive to ecology and justice concerns, what real leverage do they have in this age of economic deterritorialization and transnational capital?[16]

Policy measures increasingly provide for the environment to be "taken into account"—for example, clean air, workers' health, hazardous waste control, protection for endangered species, substitution for ozone-damaging chemicals. International treaties and agreements are established to safeguard the patrimony of humanity: a livable habitat, genetic and ecological diversity, rights for indigenous peoples and minorities. Do such actions constitute evidence that liberal democratic institutions *are* capable of responding effectively on environmental problems? And if not, do more radical (direct, decentralized, discursive) democratic principles furnish an adequate basis for response to the ecological crises in their social as well as economic dimensions? These are the two themes—critique of existing institutional forms and sketching of possibilities of radical democratic processes—taken up in different ways by John S. Dryzek; Margaret FitzSimmons and her coauthors; Colin Hay; Michael Gismondi and Mary Richardson; and Alex Demirović. Considered as a group, these essays provide an interesting spectrum of politically committed critique, although mostly confined to contexts of the industrialized North. As to the responsiveness of existing states, the general conclusion is that present political processes tend to be too straitjacketed, locked in to vested interests, and bogged down in a sort of mutual hostage taking to leave much scope for fundamental reform.

John S. Dryzek, in "Ecology and Discursive Democracy: Beyond Liberal Capitalism and the Administrative State," offers a three-pronged critical appraisal of the prevailing nexus in the Western world of capitalism, liberal democracy, and the administrative state. He argues that each mode of organization (or institutional form), considered individually, is inept when it comes to ecological concerns; and taken together they tend to compound error. However, out of the very confusion and contradictions of the state-dominated forms can arise spaces for more truly democratic action involving social movements grounded directly in (inter alia) communal, ecological, feminist, peace, health, urban space, and indigenous autonomy concerns. *Discursive* democratic forms of debate and decision making, which are open to real diversity of voices and to self-transformation, are avenues in the collective search for cooperative solutions.

The essay by Margaret FitzSimmons, Joseph Glaser, Roberto Monte Mor, Stephanie Pincetl, and Sudhir Chella Rajan on "Environmentalism and the Liberal State" covers some similar ground to Dryzek, taking a comparative look at the parameters of environmental reform politics within liberal democratic structures. Drawing on material on reform politics in the United States, Japan, the Philippines, Mexico, and Brazil, they explore the inertia and conservative biases inherent in existing "liberal" democratic and state-bureaucratic forms. The liberal state tends to straitjacket expression of environmental concern, and to channel it into forms that do not put in question the continued operations of corporate capitalism; thus the prospects for radical reforms lie more with the limits to the extent of hegemony of these state forms, and with the spaces for autonomous action opened up by their fractures and internal contradictions.

Colin Hay, in "Environmental Security and State Legitimacy," also focuses on the responsiveness of the state in Western democracies, particularly in terms of the constraints imposed by insertion in a capitalist world economy and by their conflicting obligations to capital (business) and to electoral constituencies. Ecological concerns do constitute, he proposes, a severe "legitimation crisis" to which these states must respond. But the tendency of states to engage in tactical "crisis displacement" results, most often, in patch-over responses that do not get to grips with the underlying threats (global and local) to environmental security. In practice, then, capitalist reforms (the "greening" of capital) are largely cosmetic, and state initiatives in favor of environmental quality and conservation often function as means for capital to recuperate the "crisis" to its own ends.

Looking to the future, this sort of diagnosis probably means an intensification of struggles within and between nations, and more fragmentation and alienation within societies. However, the hegemony of capital is not total, and the contradictory agendas to which "liberal" governments must respond (under fiscal constraints) leave ambiguous terrains for political action. Given this reality, can one perhaps look to possibilities of broad coalitions in search of more cooperative, autonomous, and "democratic" social forms? Several contributions to this book allude to this hope. Desperate rearguard struggle and the search for cooperation are two sides of the same "resistance" coin. These are the territories that need to be explored in theory and in practice over coming years.

Michael Gismondi and Mary Richardson give an example of politicized struggle, in their detailed analysis of "Discourse and Power in Environmental Politics: Hearings on a Bleached Kraft Pulp Mill in Alberta, Canada." Their view of the broad political process is, like John Dryzek's, one of qualified optimism. They suggest that the fracturings of reality through collision of interests can be empowering for people in opposition to the hegemonic interests of capital and the bureaucratic apparatus. But, as their example shows, the struggle is a hard one; and in a postscript they add that, while the pulp mill project was at one point blocked on environmental and worker health grounds, a revamped project has since been given the go-ahead, though with modifications to reduce some of the most severe chlorine pollution.

The final contribution, by Alex Demirović, on "Ecological Crisis and the Future of Democracy," returns us to the theoretical plane of democracy. Writing in the context of debates among German Greens on political strategy and reform, he critiques two prominent German theorists' proposals of what would constitute an ecologically sound constitutional framework. He addresses what is rapidly becoming a central question of the "sustainability–democracy" nexus: how, and in what terms, might a collective social process be created, emancipative for its participants, that simultaneously prevents the usurpation of the future by the actions of the powerful today?

3. Liberty, Democracy, and Reciprocity

Most of the essays conclude with more questions than answers. Each contribution privileges specific domains, theoretical perspectives, and modes of political response; but there are no easy formulae being laid out. This lack of prescriptions for concrete action may leave some dissatisfied. But I think it is healthy. As Serge Latouche once pointed out:

> All leaders have need of counsel, and there are people paid to give it. Ordinary people have a practical need also, to understand the world in order to achieve some degree of self-determination as a free subject; the civil servant, by contrast, merely needs a set of instructions in order to carry out his task. One of the misfortunes

of our times is that we have confused these two roles—those of the free citizen and of the functionary—to the point that understanding and liberty tend to be reduced to the instrumental level of informed choice.[17]

Environmental crisis is forcing us to rethink basic notions of what constitutes human dignity and freedom. Of course, the proponents of liberal democracy idealize elections that are "free and fair" based on "one man, one vote"; and market society proclaims itself as the incarnation of freedoms of choice. In the name of the free market, aggressively and deliberately lands and peoples are *opened up* to commodification and the intrusions of transnational capital, to the play of market forces. And who, after all, can resist the enchantment of product and life-style diversity on the world supermarket shelves— especially when looking down the barrel of a gun? But these notions of freedom are defective. Behind the kaleidescopic facade of "liberty" so much lionized in the West lies an *instrumental* logic by which nature—and also people—are reduced to the status of means to another's end.

The West can, in one specific sense, be considered emancipative, in that it frees people from the thousand constraints of "traditional society" and opens up an infinity of possibilities. Modern man, pretending himself freed from the shackles of tradition and transcendent obligations (such as to God or to the tribe), is also free to create a history of his choice. In the instrumental view of nature, the nonhuman world is considered as a freely available raw material to be put to use. To speak of energy in the 19th century was to evoke the immense powers of nature potentially at work for progress in industrial production. The (meta)physics of *liberation of energy* thus goes hand in hand with the *emancipation of the individual*, the presumed *freedom to choose* and to dispose freely of one's wealth. As Jean Baudrillard has put it:

> Our culture is witness to an irreversible process of liberation of energy. All other societies have relied on a pact of reciprocity with the world, on a stable prescription where energy doubtless entered the piece, but never a principle of liberation of energy. Man himself is, in this sense, liberated as a source of energy; and he becomes, by this, the motor of a history and of an acceleration of history.[18]

But there are fishhooks right through this liberty metaphysics. The liberal utopia, of course, is equal access for everyone to the benefits of the industrial machine. But neither the material nor the ethical basis for realization of this utopia is furnished along with the technical package. Substantive freedoms will actually be obtained only for minorities, while solidarity and security will be destroyed for everyone. In practice:

> The individualist worldview is like a yeast for the decomposition of social ties. It eats away at the tissue of traditional solidarities like a cancer. The thing that renders individualism irresistible, is that to each individual it appears as liberation. It emancipates, in effect, from constraints and opens up unlimited possibilities--but at the expense of the solidarities which constitute the fabric of social collectivity.[19]

How often has it been argued that the "magic of the marketplace" is that while each individual treats everyone else as a means to their own ends, simultaneously they are respecting the others ends-in-themselves? But this supposed respect is a thin veneer over disdain, manipulation, fraud, and neglect. Private vice does *not* sum to public virtue. We might note that the doctrine of mutual respect of individual rights and freedoms within society validates absolute freedom *provided only that* such enjoyment/disposal of one's property and self does not impair the legitimate freedoms of others. But note: the poor confront every day their "right" to count for nothing in market society; women and wage workers every day confront the fact of their "freedom" to be of service in the projects of others, which they undertake (it is said) as matters of contractual free choice; and future generations no doubt will freely accept the polluted patrimony that is passed on to them. So much for equality and fraternity. Live-and-let-live turns out to be devil-take-the-hindmost and live-and-let-die.

In fact things are worse than mere neglect. Our freedom to *take* (from nature and, opportunistically speaking, from the future) is matched by our ill-considered *liberality*. William Catton recounts the fate of a small town of Times Beach, Missouri, in the American Midwest, which was irreparably contaminated with dioxin and finally evacuated through a mass buyout of the community by the U.S. Environmental Protection Agency (EPA).[20] The dioxin had accumulated in factory sludge as a by-product of the manufacture of the antiseptic hexachlorophene. When the manufacturing plant was closed early in the 1970s, the sludge was combined with waste oil and sprayed as a dust-settling mixture on the unsealed streets of Times Beach. Health complaints over ensuring years led eventually to EPA identification of the dioxin; and then a winter flood spread the con-taminated dust/silt through the flood-plain homes. Access to the ghosttown is now barred by fences and armed guards, with huge warning signs stating "Hazardous Waste Site." Entry is possible only after signing a "General Release of Liability" by which all risks and perils are assumed by the visiting individual. Catton remarks: "Manufacturing processes will always create unwanted by-products along with desired products; increasingly our only escape from the 'by-products' will be to forgo the 'products.' "

The liberty metaphysics here has turned full circle. As Catton concludes, "We are all living in a global Times Beach."[21] By our sovereign consumption choices we collectively have signed a "General Release" on behalf of all our children and our children's children. So much for instrumental reason. Nature exacts a subtle retribution for our taking such liberties with her powers. Somehow, liberty curves into catastrophe. Today, a high-tech misadventure or "industrial accident"—particularly nuclear or genetic—has the capacity to cause life-threatening disruption *across the entire planet and permanently through time.* "I survived Three Mile Island," read a popular T-shirt a few years ago. We *all* live at Three Mile Island, at Chernobyl, at Bhopal (and, contrary to the hope expressed on the T-shirts, we may not survive).

Socially too all this freedom becomes a trap. The sovereign consumer becomes isolated in unsocial suburbia, coldly comforted by the smorgasbord of TV. The "winners" (in the "win-win" game that makes most into losers) achieve their power at the price of self-alienation from that on which their power is played. This was made apparent by the barricades dividing rich from poor at Rio de Janeiro. As Latouche observes, "Violence, insecurity, and terrorism install themselves at the doors of the rich"; but really, "all this violence attributed to the Other, which is relayed back to us as in a mirror, is only the violence that we have not ourselves been able to affront or master."[22]

Of course, capitalism portrays all these manifestations of ecological contradiction—limits to growth and limits to control, eternal scarcity, and perpetuation of misery—not as fatal limits but as challenges to be taken up: as tasks for technology and good management to overcome. But one is permitted to doubt the ultimate success of this defiant wager. It is, after all, only one way of looking at things, and a fairly one-eyed way at that. Consider the resource depletion problem. There is actually no danger that the world will run out of "free energy." The problems of exhaustion and degradation that we face are of a different sort. Pamela Zoline, in "The Heat Death of the Universe," points out that the "average housewife" understands perfectly well the meanings of increasing disorder and entropic irreversibility:

Sarah Boyle writes notes to herself all over the house; a mazed wild script larded with arrows, diagrams, pictures; graffiti on every available surface in a desperate/heroic attempt to index, record, bluff, invoke, order and placate. On the fluted and flowered white plastic lid of the diaper bin she has written in Blushing Pink Nitetime lipstick a phrase to ward off fumy ammoniac despair. "The nitrogen cycle is the vital round of organic and inorganic exchange on earth. The sweet breath of the Universe." On the wall by the washing machine are Yin and Yang signs, mandalas, and the words, "Many young wives feel trapped. It is a contemporary sociological phenomenon which may be explained in part by a gap between changing living patterns and the accommodation of social services to these patterns." Over the stove she has written, "Help, Help, Help, Help, Help."[23]

The captains of industry in the 19th century built capitalism on the strength of the steam engine. But we, the participants of this society, are the molecules *inside* the combustion chamber, rammed by the pistons, spat out the exhaust tubes. And the whole ratiocinated machine seems to have run amok. A mutually respectful coexistence on this planet would better be pursued on the basis of a generous-spirited solidarity, *relinquishing* (1) the Enlightenment control myth, and (2) the norm of self-interested accumulation. This means something is to be learned from notions of decency and reciprocation to be found in many nonindustrial cultures, and also from the wisdoms of the many subcultures within the West itself—whose contributions on economic and ecological matters have frequently been ignored or held in disdain. As both Juan Martínez Alier and Ariel Salleh have remarked in their contributions to this volume, much wisdom for the invention of ecological utopias is to be found in the South; and it is both a tragedy and a scandal that the existence of this real "egalitarian ecologism" is being ignored in most of the high-level struggles to control the world environmental agenda.

The instrumental view of nature and of human relations that permeates capitalism chooses not to heed the many voices, touches, and dramas of nature as poetry and passion; preferring rather to gauge people and things for their utility in self-interested pleasure and accumulation goals. This is an impoverished (and self-mutilating) sensibility. Hegel wrote in 1807, with a certain ecological perspicacity, that *utility* is an implacably reciprocal relationship. Materially speaking, everything depends on everything else; and

Just as everything is useful to man, so man is useful too, and his vocation is to make himself a member of the group, of use for the common good and serviceable to all. The extent to which he looks after his own interests must also be matched by the extent to which he serves others, and so far as he serves others, so far is he taking care of himself: one hand washes the other.[24]

What is in one sense a "brute" fact of coexistence may also be affirmed as an *ethic*. That is, by affirming a necessity of being in the service of the other, a destiny to be of use to the other, we may also affirm that, when resolved in a social pact of reciprocity, the deepest pleasure in life comes with this existential "vocation" of being in the service of other life. A truly materialist economic and political alternative would thus mean the abandonment of the control imperative of instrumental reason, the abandonment of the closed loops of cybernetic design, in favor of an openness in sensuous reciprocal relationship. For example, Maria Mies has suggested:

A feminist concept of labour has therefore to replace the predatory economic relationship of Man to "nature" by a co-operative one. . . . A feminist concept of labour has to reject the notion that all "necessary labour" is a burden that should be done by machines and robots. We have to maintain a concept of labour in which "enjoyment" as well as the "hardness" of work are united. This would require a different economy from the one we know today.[25]

What new forms and meanings could we give to the fact of life of each being in the service of each other, that each individual and human society be a member of the group, "of use for the common good and serviceable to all"? That is, perhaps, one of the themes that discursive democracies will have to discuss.

4. Thanks and Acknowledgments

I wish to extend thanks to all the contributors to this volume, for their efforts in writing and revision, and communicating around the world; to the various translators and manuscript reviewers who helped the process; to James O'Connor who provided valuable editorial advice and assistance; and most particularly to Barbara Laurence who undertook much of the manuscript production work. I also wish to thank The Guilford Press, and numerous friends and colleagues here at the University of Auckland and elsewhere for their support, care, and stimulus.

Notes

1. Cited in François Brune, *Le Bonheur conforme* (Paris: Gallimard, 1985), p. 233.

2. Serge Latouche, *La Planète des naufragés* (Paris: La Découverte, 1991), English translation by Martin O'Connor and Rosemary Arnoux: *In the Wake of the Affluent Society* (London: Zed Books, 1993), pp. 89–90.

3. Ilya Prigogine and Isabelle Stenders, *Order out of Chaos* (London: Heinemann, 1984), p. 111. See also the essays by Juan Martínez Alier, Jean-Paul Deléage, and Martin O'Connor in this volume.

4. David Ricardo, *On the Principles of Political Economy and Taxation*, in Piero Sraffa, ed., *Works and Correspondence* (Cambridge: Cambridge University Press, 1951), p. 69.

5. See, for example, Richard Norgaard, *Development Betrayed* (London: Routledge, 1994); Wolfgang Sachs, ed., *The Development Dictionary: A Guide to Knowledge as Power* (London: Zed Books, 1992); and Wolfgang Sachs, ed., *Global Ecology* (London: Zed Books, 1993).

6. Pamela Zoline, "The Heat Death of the Universe," in Pamela Sargent, ed., *The New Women of Wonder: Recent Science Fiction Stories by Women about Women* (New York: Vintage Books, 1978), pp. 99–120; quote from pp. 100–102.

7. Maria Mies, *Patriarchy and Accumulation on a World Scale* (London: Zed Books, 1986). She uses the term specifically to refer to women in Third World countries, who engage in labors of reproduction, in domestic and subsistence work, and (often) also in wage labor, but themselves receive less than enough to live on.

8. It goes without saying that the "alienated" wage workers have antagonistic, passified, and ironical relations to the processes that do not net much surplus for them.

9. See James O'Connor, "Capitalism, Nature, Socialism: A Theoretical Introduction," *CNS*, *1(1)*, no. 1, Fall 1988; and also his essay in this volume.

10. Ibid., pp. 32–33.

11. Giovanna Ricoveri, "Culture of the Left and Green Culture: The Challenge of the Environmental Revolution in Italy," *CNS*, *4(3)*, no. 15, September 1993, p. 117.

12. Ivan Illich discusses this sort of occulted drudgery of modern life in *Shadow Work* (London: Marion Boyars, 1973). A case in point is the labor involved in the "recycling" of containers and packaging—here again the manufacturer manages to offload costs onto consuming "households"; see, for example, Irmgard Schultz's discussion of waste disposal policies in Germany, "Women and Waste," *CNS*, *4(2)*, no. 14, June 1993.

13. See, for example, Juan Martínez Alier, "Distributional Obstacles to International Environmental Policy: The Failures at Rio and Prospects after Rio," *Environmental Values*, 2, 1993, pp. 97–124; and his "The Loss of Agricultural Biodiversity: An Example of the 'Second Contradiction,'" *CNS*, *4(3)*, no. 15, September 1993; and also see Vandana Shiva, *Biotechnology and*

the Environment (Pulau Pinang, Malaysia: Third World Network, 1992); Erna Bennett, "Whose Biodiversity?," presentation from the Rural Advancement Foundation International (FAFI) to the Fenner Conference, Canberra, Australia, July 1993.

14. This sort of commodification can be understood as, in some respects, the transition from the seed as a common property of a "gift community" to its locking up as "private property" in a liberal society; see Lewis Hyde, *The Gift: Imagination and the Erotic Life of Property* (New York: Vintage Books, 1979), especially chap. 5 on "The Gift Community."

15. Vandana Shiva, "The Seed and the Earth: Technology and Colonisation of Regeneration," unpublished manuscript, 1992, p. 31.

16. On the question of the nation-state, see Ashis Nandy, "State," in Wolfgang Sachs, ed., *Development Dictionary*, op. cit., pp. 264–274; Basil Davidson, *The Black Man's Burden: Africa and the Curse of the Nation-State* (London: James Currey, 1992); and Serge Latouche, *L'Occidentalisation du monde: Essai sur la signification, la portée et les limites de l'uniformisation planétaire* (Paris: La Découverte, 1989).

17. Latouche, preface to *In the Wake of the Affluent Society*, op cit.

18. Jean Baudrillard, *La Transparence du mal* (Paris: Galilée, 1990), p. 105; my translation.

19. Latouche, *L'Occidentalisation du monde*, op cit., p. 54 and p. 109; my translations.

20. William R. Catton, Jr., "Cargoism and Technology and the Relationship of These Concepts to Important Issues Such as Toxic Waste Disposal Siting," in Dennis Peck, ed., *Psychosocial Effects of Hazardous Toxic Waste Disposal on Communities* (Springfield, Ill.: Charles C. Thomas, 1989), pp. 99–117.

21. Ibid., p. 114.

22. Latouche, *L'Occidentalisation du monde*, op. cit., p. 56 and p. 109; my translations.

23. Pamela Zoline, "The Heat Death of the Universe," op. cit., p. 104.

24. G. W. F. Hegel, *Phenomenology of Spirit* (German original, 1807; English translation by A. V. Miller, Oxford: Clarendon Press, 1977), pp. 342–343.

25. Maria Mies, interviewed by Ariel Salleh, "Woman, Nature and the International Division of Labour," *Science as Culture*, no. 9, 1990, pp. 73–87.

Ecological Economics and Ecosocialism

Juan Martínez Alier

1. Introduction

A long line of ecological economists have seen the economy not as a circular flow of exchange value, a merry-go-round between producers and consumers, but rather as the one-way entropic throughput of energy and materials.[1] Since the economy is entropic, it is marked by exhaustion of resources and also by the production of waste. The ecological critique of mainstream economics is based on the question of unknown future agents' preferences and their inability to come to today's market, and therefore on the arbitrariness of the values given at present to exhaustible resources or to future social and environmental costs. This critique is also based on uncertainty about the workings of environmental systems, which prevents the application of externality analysis: we do not know about many externalities, and concerning some externalities about which we *do* know, we do not even know whether they are positive or negative (or *for whom* they are positive or negative in their effects). Certainly, we are far from being able to give to them a monetary present value. In sum, the ecological critique points out that because of the temporal dimension in material life, the economy involves allocations of waste and diminished resources to future generations, and moreover, such allocations arise without any transactions with these future generations, who will be

Professor of Economics and Economic History, Universidad Autónoma, Barcelona 08193, Spain. First published in *CNS, 1(2)*, no. 2, June 1989, pp. 109–122; reprinted here with minor corrections.

impacted by them. Therefore, the economy cannot be explained on the basis of individual choices and preferences. Methodological individualism encounters the insuperable ontological difficulty of coping with future generations. Because of this, ecological economics is a main enemy of orthodox economics.

To see the economy as entropic does not imply ignorance of the antientropic properties of life. This point must be made explicitly because of the growth of "social Prigoginism," that is, the doctrine that "social systems" (e.g., Japan, or the European Common Market, or the city of New York) self-organize themselves in such a way as to make worries about depletion of resources and pollution of the environment redundant. While the recognition that *physical–chemical* structures in open systems can be antientropic is relatively recent in science (the mathematical modeling of such processes dates back no more than three decades), the idea of "life against entropy" is over 100 years old. It has been very much a part of the ecological view of the economy.

Georgescu-Roegen, in his book *The Entropy Law and the Economic Process*,[2] espouses an ecological economics that clearly would be opposed to "social Prigoginism." But it would not be opposed to the view that systems that receive energy from outside (such as the Earth) may exhibit steadily increasing degrees of structure and organization over time.[3] Much earlier, Vernadsky (1863–1945) had argued, in a section of his book *La Géochimie* entitled "Energie de la matière vivante et le principe de Carnot," that the energetics of living matter were contrary to the energetics *de la matière brute*.[4] This same idea had also been pointed out by authors such as the Irish geologist John Joly and the German physicist Felix Auerbach (with his notion of *Ektropismus*). One could find the idea too in J. R. Mayer, Helmholtz, and William Thomson (Kelvin). Vernadsky added:

> The history of ideas concerning the energy of life presents an almost unbroken series of thinkers, scientists, and philosophers arriving at the same ideas more or less independently. . . . A Ukranian scientist who died young, S. Podolinsky, understood all the significance of such ideas and he tried to apply them to the study of economic phenomena.[5]

Given the importance of Vernadsky's work in the science of ecology and also in the current ecological revival in the former Soviet Union, one could say, with no disrespect, that this is an endorsement of ecological economics straight from the horse's mouth. Although Podolinsky (1850–1891) was a Darwinist, he was not a social Darwinist. In fact, he was a rabid radical federalist, a Ukrainian *narodnik* (pro-peasant socialist), a physiologist, and a Marxist: in all, a most attractive character. He attributed differences in use of energy within and between nations not to any evolutionary superiority, but rather to the inequalities bred by capitalism.

2. History of Ecological Thought in Economic Planning

It is clear that environmental concerns and reliance on the market are mutually antithetical. Therefore, one might have expected ecological issues to figure importantly in the debates on economic planning in the central European context of the period stretching from late 19th century to the 1930s. The ecological discussion was weak in socialist circles, however, because of the lack of an ecological Marxism, and also because of the absence of an ecological anarchist movement. Few authors wrote about a collectivized economy with ecological considerations in mind.

There were some remarkable exceptions, including Joseph Popper-Lynkeus (1838–1921), Karl Ballod-Atlanticus (1864–1933), Otto Neurath (1882–1945), and William Kapp (1910–1976).[6] Neurath's concept of a *Naturalrechnung* (i.e., accounting in nature) was developed because he realized that, from an ecological point of view, elements in the economy became incommensurable. Neurath's idea was received by market economists in a manner that could have been predicted: Hayek wrote that Neurath's proposal that all calculations of the central planning authorities should and could be carried out *in natura* showed that Neurath was quite oblivious of the insuperable difficulties that the absence of value calculations would put in the way of any rational economic use of resources.[7] Hayek, along with almost all participants in the debate on economic rationality under socialism, was quite oblivious to problems of resource depletion and pollution. Hayek's glorification of the market principle and individualism led him to dismiss authors who developed critiques of economics from the ecological point of view—authors such as Frederick Soddy, Lancelot Hogben, and Lewis Mumford, as well as Neurath—as totalitarian "social engineers."[8] Teachers of "comparative economic systems" have gone over and over the debate on economic calculus in a socialist economy, perhaps praising Lange's ingenious "market socialist" solution to Max Weber's, Ludwig von Mises's, and Hayek's objections, without realizing that the debate should have included, and actually did include (because of Neurath), a discussion on the intergenerational allocation of exhaustible resources. It should be stressed that this debate is different in kind from the conventional

economic question of whether coal or oil should be priced according
to marginal cost of extraction instead of average cost—as if either
method would ensure an optimal intergenerational allocation.

Neurath, inspired by Popper-Lynkeus and Ballod-Atlanticus,
was aware that the market could not place appropriate values on
intergenerational effects. In his writings on a socialist economy, starting in 1919, he put forward the following example. Two capitalist
factories, achieving the same production, one with 200 workers and
100 tons of coal, the second with 300 workers and only 40 tons of
coal, compete in the market. The one using a more "economic"
process would achieve an advantage. In a socialist economy, in order
to compare two economic plans, both of which achieve the same
result, with one using less coal and more human labor, the other
using more coal and less human labor, we would have to decide on
a present value of future needs for coal. We must, therefore, not only
decide on a rate of discount and a time horizon, but also guess the
evolution of technology (e.g., use of solar energy, water power,
nuclear power). To this we would need to add considerations concerning global warming, acid rain, and radioactive pollution. Because of
this heterogeneity, a decision about which plan to implement could
not be reached on the basis of a common unit of measurement. Elements of the economy are not commensurable, hence (according to
Neurath) the need for a *Naturalrechnung*. One can see why Neurath
became Hayek's bête noire. In fact, he managed to antagonize liberals
and moderate social democrats (because he was too radical and not
sufficiently pragmatic), as well as official Communists (because he
did not worship the Soviet experiment). Thus, he was said to think
auf so primitive chiliastische Weise that he was *im Utopismus stecken
geblieben* (he thought in such a primitive chiliastic manner that he had
remained stuck in Utopianism)![9]

As is well known, Neurath wrote in defense of "scientific utopias." He was not only an economist and a radical (active in the
revolution in Munich in 1919), but also a major analytical philosopher
of the Vienna circle, one manifesto of which he wrote himself. Many
Austrian politicointellectual battles, such as that between Hayek and
Neurath, are still relevant today.[10] Here, we reopen one of the major
polemics of our age, by pointing out that the market economy by
itself cannot provide a guide for a rational intertemporal allocation
of scarce resources and of waste.

3. Incalculable Externalities

Neurath proposed the drawing up of many "scientific utopias," such
as those of Popper-Lynkeus and Ballod-Atlanticus, from which con-

crete plans for running a socialized economy could be chosen. One can understand the nostalgia for the freedom of the market that many feel when confronted with the specter of a centralist ecobureaucracy (where *eco* could refer to "ecological" as well as "economic": a bureaucracy with the technocratic secrets of economic and/or ecological planning). The market does give freedom, even if for many it is "freedom to lose" rather than "freedom to choose." But, in intertemporal perspective, this freedom is illusory. The market economy cannot deal with ecological problems because it cannot cope at all with diachronic externalities. Future generations have no freedom to choose. Let us look at two examples that have been discussed for a long time: the increased "greenhouse effect" and nuclear power.

Svante Arrhenius explained in his textbook on global ecology[11] that the *Glashauswirkung* (i.e., the greenhouse effect) that helped to keep the Earth warm perhaps would increase somewhat with the increase in carbon dioxide (CO_2) in the atmosphere, and that in northern latitudes this change was to be welcomed. In 1937, it was estimated by a British scientist that fuel combustion had added about 150,000 million tons of CO_2 to the air in the past 50 years, three-quarters of which had remained in the atmosphere. The rate of increase in mean temperature was estimated at 0.005°C per year. Callendar remarked: "The combustion of fossil fuel . . . is likely to prove beneficial to mankind in several ways, besides the provision of heat and power. For instance, the above-mentioned small increase [0.005°C] of mean temperature would be important at the northern margin of cultivation."[12] The author quoted was, by his own description, "steam technologist to the British Electrical and Allied Industries Research Association," but his paper was received and discussed amiably by disinterested, objective scientists belonging to the Royal Meteorological Society of Great Britain. These scientists questioned Callendar's statistics but not his view that increased CO_2 would be a positive externality.

Perhaps there were discussions concerning the increased greenhouse effect by people who burned less coal in other latitudes where, in principle, a small increase in temperature would be less welcome than in Stockholm or London. Research on the sociointellectual history of climatic change up to the scare in the United States in the summer of 1988 has now become an interesting subject, and perhaps we will learn that some scientists began to take a pessimistic view of such climactic change. Whatever, if international environmental policies based on CO_2 budgets are established (either setting compulsory upper limits, or taxing emissions over a stated limit), one could well argue that they should include in each country's budget the accumulated past emissions—if not from the beginning of the Industrial Revolution, at least since the turn of this century—since the effects of fossil-fuel burning on global warming were known at that time.

The global warming scare has led to arguments in favor of nuclear power, but the economics of nuclear power also provides a good example of incalculable externalities. We must assign present values to the unknown future costs of dismantling power stations in a few decades and also to the unknown costs of keeping radioactive waste under control for thousands of years, and such values will depend on the rate of discount chosen. Moreover, for some by-products of nuclear power (such as plutonium), we do not know whether to classify them as positive or negative externalities, let alone how to attribute a monetary value to them.

Since the plutonium produced as a by-product of the nuclear civil program may have a military use, it can be given a positive value, thus improving the economics of nuclear power (in the chrematistic sense). This "plutonium credit" was factored into the accounts of the initial British nuclear power stations.[13] However, plutonium might come to be seen in the future as a negative externality. Frederick Soddy, a highly qualified nuclear scientist, warned against the "peaceful" use of nuclear energy in 1947 because of "the virtual impossibility of preventing the use of non-fission products of the pile, such as plutonium, for war purposes."[14]

These examples suggest that conventional environmental economics are rather useless as a instrument of environmental management. The use of the concept of "externalities" merely reveals the economists' inability to put a value on future, uncertain, even unknown effects, whether they pertain to depletion or pollution. In any case, the economists typically ignore the role of social movements to force capitalist enterprises to internalize externalities, and all the uncertainties pertaining to these movements.

4. Limitations of Ecological Planning

The market alone cannot deal with externalities, but a purely ecological approach also has limitations.[15] Ecologists, and environmental scientists in general, are increasingly being asked to determine standards for human life, instead of pinpointing trade-offs. Examples include "safe" doses of radiation and pesticides, tolerable "CO_2 budgets," and even acceptable densities of population (at least in poor countries). Scientists are also asked to deliver new materials and new genetically engineered plants and animals, with no nasty environmental side effects. But scientists have no methodology for getting a common standard of measurement that will provide a guide to the trade-offs that are really in question.

We cannot compare in commensurable terms such costs and benefits, because of uncertainties (which are not risks with known probability distributions), and because of the moral issue of giving values to future effects. However, since economics does not provide a guide, we might be tempted to become managerial ecologists, and attempt to base decisions on an ecological rationality. Ecology and economics again conflict in the definition of "degradation of the resource base." Economists would tend to claim that the use of resources, even if they are not produced but merely extracted, is not necessarily economic degradation because before they are exhausted new resources will be substituted for them. Economists would also point out that, although there is no guarantee of such substitution, we should still use resources because the growth of the economy makes future consumption at the margin less valuable than today's consumption. A strict conservationist posture, which would give at least equal values to future versus present consumption, would perhaps lead to resources being left unused when techniques change, for example, when we have fusion energy and before we run out of oil and gas. One important element in this discussion is the rate of discount. The ecologists can always point out, with reason, that the economists have no strong arguments to impose a particular rate of discount. They could even argue for a negative rate of discount.

Nevertheless, precisely because of uncertainties about the future and the inevitability of choices being made between differing social, species, and ecosystem options, a so-called ecological rationality is not an indisputably better base for policy decisions than the usual economic rationality.

This can be illustrated by reference to the notion of "carrying capacity." Carrying capacity refers to the maximum population of a given species that can be supported indefinitely in a given territory, without a degradation of the resource base that would diminish this population in the future. Attempts to apply the notion of carrying capacity to humans, for poor countries, are often made by international agencies and the multilateral lending banks. But one could well imagine the response from European authorities if scientists from the third world were to point out that, first, Europe has exceeded its own carrying capacity, and, second, that the world's carrying capacity would be increased by removing all barriers to international and intercontinental migrations. Western Europe has a policy of restricting inmigration and of increasing or maintaining the population through a higher birth rate. The push for a higher population is based on an implicit assumption either of decoupling economic growth from the use of energy and materials (by means of increased efficiencies and recycling) or of a continuing ability to extract energy and materials at a cheap price from overseas countries in the characteristic European pattern of *Raubwirtschaft*.

Recently, there was an accident in the Mediterranean similar to those that frequently occur between Haiti and the United States. In March 1989, 10 Moroccan would-be immigrants died at sea while attempting to reach Spain. The right to choose one's place of habitation on earth remains the most elusive of human rights. On the same day, by chance, it was announced that Spain, in keeping with the notion of a "fortress Europe" for 1992, would require visas for all Moroccan, Algerian, and Tunisian travelers, as well as for Latin Americans, starting in March 1990. A government official (in his role as a sort of Maxwell's demon) explained that Spain has a long coast near countries with population problems.[16] Thus, migration and the prohibition of migration are not seen as a function of the difference in standards of living, but as the consequence of the pressure of population on resources in the South. Nevertheless, when some European nations were countries of outmigration not so long ago, their population densities (perhaps with the exception of Portugal) were lower than they are today. Migration usually is a result of "pull" factors, and in any case carrying capacity can be increased, if not from domestic exhaustible resources, then from energy and materials subsidies from outside. Across states, frontier police stop migrants who come from territories where they are not necessarily starving, but where they face a comparatively low level of consumption. States, frontiers, and policement are products of social conflicts. Hence, the Maxwell's demon analogy: Maxwell's demons were unnatural beings who were supposed to be able to maintain, or even increase, the difference in temperature between communicating gases by sorting out high-speed and low-speed molecules.

The point I wish to make is that ecologists are unable, in biophysical terms alone, to explain the territorial–political distribution of the human population. Sometimes they take the status quo for granted and then preach social Darwinism, as in Hardin's "lifeboat ethics." But, in truth, humankind's ability to maintain enormous differences in the exosomatic consumption of energy and material resources, between states and inside each state, requires distinctly human institutions. Because of the lack of genetic instructions concerning human exosomatic consumption of energy and material resources, and also because of particular political, social, and territorial human arrangements, human ecology is different from the ecology of plants and the ecology of other animals. It is a type of study that cannot be reduced to the natural sciences. Policy prescriptions based on ecological analysis make sense only if concrete social contexts are taken into account. These prescriptions cannot be more rational than the social contexts themselves. An ecological point of view may lead to social Darwinist views (applying Boltzmann's dictum of 1886, "The struggle for life is a struggle for available energy"), or equally well to egalitarian perspectives.

5. Conclusion: Economics, Ecology, or Politics?

Strong, rational arguments can be brought against both so-called economic rationality and ecological managerialism. Meanwhile, at present, there is a big struggle, fought with unequal means and unequal opportunities, to set the environmental–economic agenda in the world, and especially to determine which are the important issues. The fight is not yet about which decisions to take; rather, it is about inclusions and exclusions of topics to be discussed. Who should set the environmental–economic agenda? The government of particular states, the European Community, the World Health Organization, Greenpeace, the International Monetary Fund and the World Bank, the Chipko movement, the German Greens, the World Information Service on Energy? All parties attempt to direct the ecological debate in particular directions, but they have unequal access to the media and unequal power and money.[17]

Which territorial–political units should decide how we should live, how we should treat fellow humans, how we should treat future generations? Are the proper units of decision, regions, states, or global units? How could we combine plenary-meeting ecoregionalism and ecoglobalism? Plenary-meeting ecoglobalism seems an impractical dream, which even if brought to fruition would not guarantee proper regard for future generations. The temptation of international ecomanagerialism is strong (e.g., Harich's nightmare of "Babeuf and the Club of Rome," or even worse the "International Monetary Fund of Ecology"). Some try to escape this by praising the market, some by imagining small-scale ecoregionalist refuges, little "ecotopian" communes (presumably protected from large-scale immigration by an armed border patrol). But these visions fail to address the important questions of global ecology. After all, localized human actions may lead to irreversible ecological changes, a possibility that has long been envisaged. The gradual exhaustion of some fossil fuels caused by the great demand in a few greedy countries, species extinction because of tropical deforestation, the CO_2 buildup in the atmosphere and its hypothetical effects on climate change, acid rain, accidents in nuclear power plants, and the absence of a technical solution for the disposal of radioactive waste—all such issues were discussed at least 50 years ago, and some, 100 years ago. Lack of awareness of them is "socially constructed ignorance."[18] Other environmental effects might be genuinely surprising, for instance, the effects of chlorofluorocarbons on the ozone layer were apparently not known until a few years ago. Similarly, the awareness that a small nuclear war would be followed by a terrible "nuclear winter" or the alarm at the possible ecological effects of genetically engineered organisms are not old.

All such effects, whether noticed (and ignored) long ago or first recognized only a few years ago, are not amenable to standard externality analysis. There is much uncertainty about such effects, at least with respect to the speed at which they are happening, and also with respect to countervailing technical solutions. Economists are unable to put a present value on such effects, weighted by the probability of occurrence, and duly discounted (at which rate?).

The inability of economics to digest the ecological side effects of production and consumption, particularly the intergenerational side effects, has been pointed out by many authors. For instance, William Kapp wrote that "it is important to keep in mind that we are dealing with essentially heterogeneous magnitudes and quantities for which there can be no common denominator . . . a commensurability which simply does not exist."[19] Such warnings were repressed, such critiques unheeded, because it was politically inconvenient to acknowledge them.

Economists would perhaps retort, with some truth, that economic valuation of externalities is certainly impossible as long as the scientists themselves are unable to provide reliable physical data even on current effects, let alone on future effects. There is no doubt that the science of global ecology provides an insecure base for policies-from-above. Very likely this will always be so, and the economists will always be left forlorn. But this does not stop scientists and politicians from telling other people what they should do.

Faced with this (pseudo)scientific prepotence, it is not surprising that grassroots ecological movements turn sometimes against science. But this, in my view, is like turning against rationality. The geography of antiscientific "ecology" is peculiar: there is more of it in California than in Germany, and more in Germany than in India.[20] My own sympathies lie with the slogan of Piotr Lavrov's *narodniki*: "[Scientific] Knowledge and Revolution."

Rational discourse (in favor of science or anything else) is rather a waste of time if it is directed against enemies of rationality. Nevertheless, the point should be made that antiscientific activists sometimes conflate science and technology. Rightly mistrustful of unfounded scientific advice, aware of the deplorable ecological and social consequences of some technologies, grassroots ecologists sometimes turn against science per se, and get lost in the mists of irrationality (since they no longer possess a steadying peasant or tribal understanding of nature).

To conflate science and technology is wrong, but the antiscientific ecologists are not the only ones guilty of this sin. Authors of other political persuasions were enthusiastic about the "scientific–technical revolution" a few years ago, and this school still has many members. To mistake "scientific progress" (the advance in scientific knowledge, which undoubtedly takes place) and "technical progress" causes great confusion in grassroots ecological movements, because reasonable doubts about whether some technologies really mean "progress" become silly doubts about whether science is the right way to pursue knowledge.[21] In fact, ecological knowledge cannot be but scientific knowledge.

The progress of science has often shown technologies to be impossible or noxious. For example, ecological worries about scarcity and the allocation of energy would be unnecessary if suitable perpetual motion engines became available. Scientific progress, in the form of the laws of thermodynamics, showed this outcome to be an illusion, an impractical utopia. Although nuclear technology grew out of a marriage of science and politics, science shows why radioactive waste is dangerous and why safe storage is not possible. More recently, science has provided a critique of current agricultural practices in the overdeveloped countries, explaining why "organic farming" is superior in terms of energy efficiency and also in terms of pollution effects. Science not only provides the data in order to denounce some technologies as impossible, dangerous, or wasteful, it also concludes that, because of incomplete data and complex interactions, science itself cannot dispel uncertainties about some global environmental effects. The ozone layer issue is probably exceptional in that scientific data has been the base for consensual policies-from-above. In contrast, the valuation of the greenhouse effect is embedded in uncertainties and distributional conflicts.

Perhaps science cannot (yet?) give convincing estimates about "global warming," or about "global chilling," or about many other important ecological questions. But even when data overwhelmingly prove some ecological fact, science by itself cannot be the base for policies-from-above because science cannot guide the trade-offs implied by intergenerational allocations of resources and waste. That is, even if reliable scientific information were available, a purely science-based policy from the top down would be impossible by itself, because good ecological data are not a guide for decisions regarding distribution between different social groups and generations.

Ecotechnocratic and neoimperial arrogance have come together in many recent international conferences on the environment. Thus, a member of the new international environment establishment, the West German Minister for the Environment, was reported (perhaps unreliably) in 1989 to have proposed "a worldwide convention on the protection of the global climate, with specific limits on the production of carbon dioxide and methane, the major contributors to the greenhouse effect."[22] This foreshadowed one of the major "agendas" of the United Nations Conference on Environment and Development Rio conference. Methane, indeed! The methane (CH_4) that goes into the atmosphere and makes a small contribution to global warming comes from wet rice fields, dung, biomas burning, and natural gas. Such a proposal, as it is reported, will lead to a discussion of the production of paddy rice, conveniently steering attention away from the disproportionate contribution to resource depletion and pollution made by the rich peoples of the world. The new environmental agenda which "the International Monetary Fund of Ecology" tries to promote universally preaches "adjustments" for everybody. There is an organized attack against egalitarian ecologism in many international conferences, backed by much money and publicity, where the "rice eaters" are, of course, absent and unrepresented.

So the question of whether policies are to be based on economics *or* on the science of ecology is not the relevant one. Instead, the true questions are: Should issues of distribution be decided by the market, with unequal purchasing power and in absence of future generations? By policies-from-above, based on technocratic economics or on technocratic ecology? Or are they to be decided by a politics of universal, and much more equal, representation? This third option would be helped by drawing up many different scenarios, well informed by science, and with explicit acknowledgment of future generations: a collection of concrete, ecological, scientific utopias in the line of Popper-Lynkeus, Ballod-Atlanticus, and Neurath. Many of these utopias should come from the poor. The present tragedy is that the existence of an egalitarian ecologism in the South is being ignored in the current, fierce struggle for the world environmental agenda.

Notes

1. These include Nicholas Georgescu-Roegen, Kenneth Boulding, Frederick Soddy, Patrick Geddes, Joseph Popper-Lynkeus, Sergei Podolinsky, and Herman Daly. Ecological economics has a long unacknowledged lineage including a line of counterattacks (e.g., Max Weber's critique of Wilhelm Ostwald in 1909) and also Hayek's remarks against "social engineers" (F. A. von Hayek, *The Counter-Revolution of Science* (Glencoe, Ill.: Free Press, 1952). This lineage and these counterattacks are summarized in detail in Juan Martínez Alier, with Klaus Schluepmann, *Ecological Economics* (Oxford: Basil Blackwell, 1987).

2. Nicholas Georgescu-Roegen, *The Entropy Law and the Economic Process* (Cambridge, Mass.: Harvard University Press, 1971).

3. Jacques Grinevald, "Vernadsky and Lotka as Sources for Georgescu-Roegen's Bioeconomics." Vienna Centre Conference on Economics and Ecology, Barcelona, Spain, 1987.

4. Vladimir Vernadsky, *La Géochimie* (Paris: Felix Alcan, 1924).

5. Ibid., pp. 334–335.

6. Joseph Popper-Lynkeus, *Die allgemeine Naerpflicht als Loesung der sozialen Frage* (Dresden: Reissner, 1912) and *Mein Leben und Wirken: eine Selbsdarstellung* (Dresden: Reissner, 1924); Otto Neurath, *Wirtschaftsplan and Naturalrechnung* (Berlin: Laub, 1925); Karl Ballod-Atlanticus, *Produktion und Konsum im Sozialstaat* (Stuttgart: Dietz, 1989) and *Der Zukunftsstaat-Wirtschaftstechnisches Ideal und Volkswirtschaftliche Wirklichkeit*, 4th ed. (Berlin: Laub, 1927); William Kapp, *Social Costs, Economic Development, and Environmental Disruption*, John E. Ullman, ed. (Lanham, Md.: University Press of America, 1983). Kapp would become in the 1950s and 1960s one of the best-known ecological economists.

7. F. A. von Hayek, ed., *Collectivist Economic Planning* (London: Routledge, 1935), pp. 30–31.

8. Hayek, *Counter-Revolution of Science*, op cit.

9. Felix Weil, review of Otto Neurath (1925), in *Archiv fuer die Geschichte des Sozialismus, 12,* Carl Gruenberg, ed. (1926) (Graz: reprinted by Syndicat, 1979), p. 457.

10. See Erwim Weissel, *Die Ohnmacht des Sieges* (Vienna: Europaverlag, 1976); Friedrich Stadler, *Vom Positivismus zur "Wissenschaftlichen Weltauffassung"* (Vienna/Munich: Loecker, 1982).

11. Svante Arrhenius, *Lehrbuch der kosmischen Physik* (Leipzig: Hirzel, 1903), p. 171.

12. G. S. Callendar, "The Artificial Production of Carbon Dioxide and Its Influence on Temperature," *Quarterly Journal of the Royal Meteorological Society, 64,* 1938, pp. 223–236.

13. J. W. Jeffery, "The Collapse of Nuclear Economics," *Ecologist, 18,* no. 1, 1988.

14. Frederick Soddy, *Atomic Energy for the Future* (London: Constitutional Research Association, 1947), p. 12. This worrisome fact did not reach

the public until the 1970s in most countries because the ecological agenda, and ecological awareness, depends on political power. It is still not widely appreciated that plutonium is an extremely toxic substance quite apart from its radioactive qualities.

15. By "purely ecological" is meant decisions based essentially on biophysical considerations of, for example, energy flows, nutrition, water, species or habitat conservation.

16. *El Pais Semanal*, March 13, 1989, p. 14.

17. Although the last couple of years have seen a high profile for "environmental diplomacy," the Rio 1992 protocols on climate change and biodiversity simply mask these ongoing struggles.

18. J. R. Ravetz, "Usable Knowledge, Usable Ignorance: Incomplete Science with Policy Implications," in W. C. Clark and R. E. Munn, eds., *Sustainable Development of the Biosphere*, International Institute for Applied Systems Analysis (Cambridge: Cambridge University Press, 1986).

19. Kapp, *Social Costs*, op. cit., p. 37.

20. Ramachandra Guha, "Ideological Trends in Indian Environmentalism," *Economic and Political Weekly*, 23, no. 49, December 3, 1988.

21. This implies a broadened view of science including, for example, much peasant knowledge about plants and ecosystems.

22. *New York Times*, May 3, 1989.

2

Eco-Marxist Critique
of Political Economy

Jean-Paul Deléage

1. Introduction

A particular view of nature gradually gained currency after the Industrial Revolution. It conceived of nature as a strictly physical and
biological reality that could be tapped by technology according to
purely economic or even monetary criteria. In contrast to this view,
many natural scientists built theories making the laws of nature the
ultima ratio of human history, in complete disregard of the specifically
social aspect of the evolution of human society.

Both standpoints are partial and untenable when considered from
the historical perspective of the long-term, double-edged crisis into
which the world has entered—a crisis at once economic and ecological. Moving beyond the limits of both these approaches and adopting
the "standpoint of the totality" is the only methodological choice that
can serve as a serious basis for an analysis of the relationship of society
to nature.[1] The specific purpose of this chapter is to pose this problem,
and to submit for consideration a few of the issues raised by the
identifiable complementarities of the Marxist and bioeconomical critiques of political economy.[2]

Professor at the Université de Paris, and editor of *Écologie Politique*. Translation by
John Barzman. The author wishes to thank Andy Szasz and other *CNS* editors. First
published in *CNS*, *1(3)*, no. 3, November 1989, pp. 15–31.

2. Nature as Tap and Sink, or as First and Last Phase of Economic Activity

From its start capitalism has treated nature as unlimited. This attitude was incorporated into the theory of capitalistic practice, bourgeois economics. The economists of the last century perceived the appropriation of natural resources in terms of rent. They focused on the problem of land because, in practice, at that time, it was the main scarce natural good. As David Ricardo wrote, "The brewer, the distiller, the dyer, make incessant use of the air and water for the production of their commodities; but as the supply is boundless, they bear no price. If all the land had the same properties, if it were unlimited in quantity, and uniform in quality, no charge could be made for its use."[3] Today this way of posing problems has been decisively revealed to be untenable, given the magnitude of the resources consumed by humans and of the changes in the terrestrial environment caused by human activities. Taking the geoecological dimension of all economic processes into account is no longer only a theoretical requirement: the very safety of our planet and the vital interests of humanity depend on it.

With capitalism, a new, historically unprecedented relation was established between the economic process and nature. With the rise of the Industrial Revolution, Adam Smith noted, "Every man . . . lives by exchanging, or becomes in some measure a merchant, and the society itself grows to be what is properly a commercial society."[4] It was in this mercantile behavior, the product of the division of labor, that Adam Smith located the immense advantage of the new society. Marx, in turn, implicitly criticized this type of dependency: "The mutual and universal dependence of individuals who remain indifferent to one another constitutes the social network that binds them together. This social coherence is expressed in *exchange value*, in which alone each individual's activity or his product becomes an activity or a product for him."[5] Since Adam Smith's time, classical and neoclassical political economy have developed as a theory of exchange value, representing the economy as a closed system, whose coherence is guaranteed by the link of exchange alone. The economy is therefore represented in the form of oriented flows of services and goods, compensated by financial flows in the opposite direction. Economic analysis ends precisely where the flows of money stop. The many criticisms of this representation of economic reality are well known: the goods and services produced by human activity only appear in the economic system insofar as they exist in the form of commodities, and they drop out of sight as soon as they lose this quality.

How do the practices of production and consumption combine with the biophysical world? In bioeconomic terms, the extraction of raw materials and their elimination in the form of waste constitute the first and last phases of all economic activity. In their natural state, raw materials appear in the form of more-or-less concentrated stocks or flows. As a general rule, these resources are scattered in low concentrations. This is true of biological resources, whose distribution around the globe is such that agriculture, fisheries, and forestry are profitable only in some regions (e.g., 99% of the production of fisheries comes from the ocean's coastal rim, which covers only 10% of the ocean's surface). The same is true of minerals and fossil fuels (e.g., the earth's crystalline rock only holds an average 50 grams of copper per ton, yet the exploitation of copper is profitable today only at 100 times higher concentrations, and sites with such high concentrations are rare).

The degree of concentration of resources—as measured by geology or ecology—is a fundamental variable for their economic use because it determines to a great extent the quantity of labor necessary for their transformation from resource to raw material. "The labor expended by nature" to achieve this concentration can be quantified since it stands in inverse ratio to the energy necessary to extract the resource from the environment. For example, when one shifts from copper extracted from porphyritic ores at a 1% concentration, to 0.5% and then to 0.3% ore, the energy cost of a ton of metal increases from 22,500 kilowatts to 43,000 and 90,000 kilowatts per ton of copper, respectively. The energy cost of the extraction of 1 ton of copper from sea water, which is often presented as an inexhaustible reservoir, is estimated today at 560,000 kilowatts per ton of metal. This increase of costs as concentration is lowered is found in most fields of mineral prospection. For example, half a century ago, over 10 times more oil was discovered per meter drilled than today; the cost of an exploration well of 30,000 feet is 120 times higher than that of a well of 5,000 feet.[6]

At the other end of productive activity, the costs of recycling waste increase insofar as human activity must assume the recycling functions no longer assumed by nature. Classical examples include the rising costs of water treatment stations and facilities for reducing air pollution. But most spectacular today is the rising cost of nuclear waste, particularly actinides, which are heavy elements with a long half-life. The best known in plutonium 239, which has a half-life of 24,600 years, and which is particularly toxic, emitting alpha particles that lodge mainly in bones. In storage, such wastes are noticeably active even after several hundred thousand years. The French nuclear power program alone will produce some 100 tons of plutonium annually by the end of this century. If 99% of these materials were recovered (truly a technological feat), 1 ton would still remain in the residue after processing. After considering the possibility of recycling these actinides in breeder reactors—a line of reactors that has turned into a technical and economic fiasco—the French now plan to store them in deep geological layers, despite the warnings of many geologists who emphasize the need to be able to retrieve and reprocess these wastes materials in an undetermined future. Whatever the solution adopted, it is clear that we now have the obligation to create "strict surveillance zones" for centuries, and in the case of actinide-rich waste storage, for several hundred thousand years.

No economic theory has had anything to say about this issue. No economic model can integrate the costs of this time bomb launched into the future.

Also, agriculture has been transformed into a genuine mining enterprise, particularly in recently industrialized countries. Increase in crop yields registered since the 1950s can be explained by an even more rapid increase in energy inputs as a result of mechanization, as well as the massive use of fertilizer and sophisticated conditioning of the crops. Agriculture in industrialized countries is now totally dependent on oil products. Loss of energy efficiency is impressive when one takes into account the entire agricultural and food industry. In Britain, 6.5 calories of fossil fuel were expended for every calorie of food in 1963; 6.1 for one in France in 1975; and 9.6 for one in the United States in 1970. According to D. Pimental and M. Pimental, 16.7% of the energy consumed in the United States in the early 1980s was used in the agricultural and food-processing circuit.[7] Even when measured in terms of labor time, the supposed efficiency of the American system is illusory. Critics have already noted how ludicrous it is to assert that "an American farmer feeds 38 people, since in reality 20 percent of the American population works in one way or another to produce food, mainly in the related industrial occupations."[8]

Paradoxically, in industrialized countries, the spectacular drop in agriculture's real energy efficiency has been associated with major gains in yield per hectare and in labor productivity. In economic terms, energy is now being substituted for cultivated area, and in ecological terms, fossil energy is being substituted for solar energy. In the industrialized countries the use of fossil energy for food needs alone generally stands between 0.6 ton of energy per person per year in France and 1.1 tons of energy per person in the United States. It is strictly impossible to generalize around the world the norms of production now prevailing in North America. Such an extension would imply the allocation of almost the entire world energy output to food production alone.

To these considerations in the field of energy must be added a more specifically ecological remark. There has been constant decline in the vegetal and animal genetic potential involved in the production of our food. Modern cultivation and herding practices tend toward genetic uniformity, whereas it is diversity that confers strength to a species and preserves the future capital of agriculture.

Most problems accumulate in the final phase of the productive cycle, in the form of waste. This applies, for example, to fertilizer, particularly to nitrates no longer held down by the colloids of the vegetal soil, but instead carried away by running water. This irrationality has already led to genuine ecological catastrophes in certain regions of Europe where intensive agriculture is practiced. Thus, in late May 1988 the North Sea, from the southern shores of Norway and Sweden to the northern shores of Denmark, was invaded across some 7.5 million hectares by a sudden proliferation of the seaweed *Chrysochromulina polylepis*, which destroyed all other forms of life to a depth of 10 meters below the surface of the ocean. The cause of this ecological catastrophe was the saturation of the seawater with nutrients, particularly nitrates used by farmers of regions adjoining the North Sea, 50% of which are carried to the sea by rain and rivers.[9] One must add multiple accidents of various kinds registered downstream of the estuaries of rivers flowing through regions of intensive agriculture. Such accidents occur every year in France along the shores of Brittany. Across the Atlantic, in the estuary of the Saint Lawrence River, a proliferation of diatoms led to three deaths and hundreds of cases of food poisoning in December 1987.

The problem is therefore twofold: it arises first when the raw materials are extracted and later again when they are unloaded into the environment in the form of waste. As more accessible and concentrated deposit sites are exhausted, either larger amounts of energy have to be mobilized to tap the less profitable deposits, or new techniques must be developed, requiring a growing effort in innovation. This rising-cost model concerns the extraction of fossil fuels as well as of nonenergy resources, fishing activities as well as cereals production. As nature proves unable to recycle the new waste, similar increases develop in the cost of waste management, for example, the storage of chemical and nuclear wastes and depollution of water and air. Yet, not so long ago, Ricardo could still write about these two natural agents as "inexhaustible," so "everyone could use them." States have responded by promoting environmental policies based on neoclassical economic principles; these strive to attribute a negative value to the damage and to contain and limit such damage rather than to influence the social and economic processes from which they stem.

3. Energy and the Forces of Production

Bioeconomists view economic processes from the point of view of the principles of thermodynamics, insisting that these principles apply both to natural systems and to systems rearranged or transformed by man. The second law of thermodynamics highlights a key aspect of all productive processes: economic activity, intended to satisfy human needs, runs against the general tendency of the universe to move toward a state of greater disorder, of higher entropy. Human labor runs against this tendency toward increasing disorder of the physical world. It sets into motion the energy sleeping within nature, converts "wild" energy into "domesticated," useful energy. But to make this useful energy available, a certain amount of human energy must be expended, either in the form of energy stored in machines or in the form of living human labor.

By definition, the overall increase of entropy associated with any process of production is always greater than the local decrease of entropy corresponding to this process. Economic activity therefore does not escape the laws of physics: the organizational status of the economy only increases insofar as that of the universe as a whole decreases.[10] As Nicholas Georgescu-Roegen observes, "In the perspective of entropy, every action of a human or of an organism, and even every process of nature, can lead only to a deficit for the overall system." "Not only," he continues, "does the entropy of the environment increase with every liter of gasoline in the tank of your car, but a substantial part of the free energy contained in the gasoline, instead of driving your car, will be reflected in a further increase of entropy. . . . When we produce a copper sheet from copper ore, we reduce the disorder entropy of the ore, but only at the price of a further increase of entropy in the rest of the universe." Living beings too are subject to this law. Every living organism, including human beings, strives to maintain its own entropy at a constant level by drawing low-entropy energy from its environment, particularly in the form of food. According to Georgescu-Roegen, "In terms of entropy, the cost of any economic or biological undertaking is always greater than the product, in such a way that activities are necessarily reflected in thermodynamic deficit."

Labor is not, all on its own, the primary self-renewing power conceived by Marxist theory. Its reproduction depends totally on a continuous input of low-entropy energy. This energy derives from the sun directly (rays, heat) or indirectly (wind, hydraulics), from solar radiation stored in fossil fuels (oil, coal, gas), and, in small part, from geothermal flows and nuclear energy. Energy cannot be created by labor or machines: it is always drawn from the environment. Even this extraction is governed by certain constraints. Just as labor is necessary to produce labor, energy is necessary to extract energy from the environment. And just as in a growth economy labor can produce more than what is necessary for its own reproduction, so the energy extracted from nature is generally greater than the energy expended for its extraction. The ratio of labor obtained to labor expended is a critical magnitude in economics: it is imperative that it be greater than 1. Similarly, the surplus corresponding to the difference between energy obtained and energy invested is net energy.[11]

Energy is therefore an essential determinant of the development of the productive forces. *At bottom, to produce is to metabolize natural physical energy into energy useful to man.* The social value of the energy freed in the form of products is obviously not identical to its physical quantity, but it is quite closely related to the amount of energy expended to obtain it. But this relation constitutes only one aspect of real social value. Another aspect is the productivity of human labor, which stems from its biological ability to tap and convert natural energy. This ability is not a mechanical but an informational quality. It is through a flow of information, through the elaboration of a human order, that society can tap the energy of the cosmos, or that humans can become "the masters and owners of nature." In short, no useful, domesticated energy is derived from wild energies without a chain of converters. Without the early plough, the earth would have been as sterile as the wind without a windmill, or uranium without a nuclear reactor.

In modern industrial economies abiotic sources of energy have overwhelmingly replaced not only human labor but also all the resources of biological origin. Thus, biological energies (wood, vegetal resources, animal traction), which accounted for one-third of the total energy consumed in France at the end of World War II, today account for less than 5% of the energy balance sheet. The share of human energy has dropped from 7% of total energy at that time to well below 1% today.

In the United States human labor represented less than 1% of the mechanical energy used in industry in the early 1980s.[12] Empirical analyses as well as historical data confirm the existence of a close connection between the quantity of fossil energy consumed, the state of economic activity, and overall economic output. This connection is independent of the nature of the economic system. Smaller-scale observations have also noted a correlation between the consumption of energy per hourly unit of labor and the productivity of labor. Cleveland has studied this correlation over a long period (1909 to 1981) in U.S. history.[13] The same is true of agricultural production over very long periods.[14] The progress of technology is obviously the essential mediation in these changes, for it increases the productivity of labor and makes it possible to move machines by a motor force independent of human energy. As Marx noted more than 100 years ago:

> Just as the individual machine retains a dwarfish character, so long as it is worked by the power of man alone, and just as no system of machinery could be properly developed before the steam engine took the place of earlier motive powers, animals, wind and even water; so too, modern industry was crippled in its complete development, so long as its characteristic instrument of production, the machine, owed its existence to personal strength and personal skill, and depended on the muscular development, the keenness of sight, and the cunning of hand.[15]

Of course, social factors determine the quantity of energy allocated to the productive effort as well as the share of this effort extracted from workers in the form of surplus value. Nevertheless, it remains true that it is the net amount of energy extracted from the various sources that constitutes the physical matrix of all economic activity.

4. Capitalism

Rates of resource utilization and waste release (i.e., of the ecological cost of production) suddenly increased with the advent of industrial capitalism. This burst was the consequence of the revolution introduced by capitalism in terms of the goals of production. According to Marx, in all previous modes of production, "the repetition or renewal of the act of selling in order to buy is kept within bounds by the very object it aims at, namely, consumption or the satisfaction of definite wants, an aim that lies altogether outside the sphere of circulation." With the advent of capitalism, "when we buy in order to sell, we, on the contrary, begin and end with the same thing, money, exchange-value; and thereby the movement becomes interminable."[16] To the removal of the limit on the circulation of capital corresponds a sudden increase in the immediate and cumulative rates of transformation of resources into waste. In my view, this is the secret of productivism, or production for production's sake, the favorite target of the ecological movement's critique of industrial society. It was the capitalist mode of production that accomplished the "unification of differences" by combining them inside a worldwide system that is at once diversified, hierarchical, and centralized (i.e., a single market) and a set of poorly adjusted regional markets.

This unification everywhere has brought about a double-edged ecological and social crisis. This applies to industrial activities that—when they are highly polluting and require a regimented work force—are increasingly relocated in the Third World. But this double-edged crisis has an even more concentrated impact on the peasantries of the entire world. Today the gap in the productivity of agricultural labor in the periphery versus agricultural labor in the center is on the average as 1:100, and sometimes wider in the most extreme cases. Yet the unification of the world market has meant the unification of world prices for all basic products (cereals, meat, etc.). The consequence is unavoidable. Less productive farmers are eliminated, whether they belong to the less favored regions of the center or of the periphery. This mechanism of destruction of the most fragile peasantries, which has occurred several times in the history of capitalism, is today wreaking havoc on a planetary scale with a violence unprecedented in history. One billion peasants have been ruined; that is, the equivalent of all manual nonirrigated farming and all farming by light animal-drawn implements has been eliminated. This billion people can now be found in the cities where they take part in the informal economy, which is characterized by very low productivity. This massive crisis of the peasant economy knows no borders; its consequences are felt throughout the whole world economy as it spreads mass poverty in the Third World. Meanwhile, in the center, as limits to the rate of exploitation of labor power are encountered, the ultimate recourse is often a frenzied destruction of the capital embodied in nature. As Lemechev remarked, "Economic growth located at the surface of social development is in fact only a Pyrrhic victory as it is obtained only at the price of the destruction of the environment and the degradation of the human habitat."[17]

5. Existing Socialism

It would be reductionist to equate the logic leading to the destruction of the environment with the logic of private profit. Societies that have abolished or statized private profit have not escaped the most brutal dimensions of the ecological crisis. Political ecology is thus now the focus of a public and impassioned discussion in the former Soviet Union and most East European countries. This should surprise only the "old believers of socialism," those who think that these societies have "achieved socialism," or at least begun its construction. In fact, their productive system is, when considered in terms of its methods and goals, too little different from that of the developed capitalist countries to avoid reproducing on a large scale the waste associated with the economic development of the latter. Whether guided by private capitalists or by the imperatives of state capital, industrialization, and the accumulation and concentration of capital that industrialization implies, have a logic of their own, identical in both cases, formally unique and independent of any theory or ideology.

Whether statized or not, industrialized economies operate under the same natural, ecological, and thermodynamic constraints. In existing socialist societies, eliminating the mechanisms of private profit has, in fact, only resulted in state control. In either case, one has to recognize that any society's capacity for labor is bounded by the global level of net energy available to it, limits imposed by technology *and* nature. State ownership of the means of production provides no theoretical guarantee against the waste of resources generated by the anarchy of the capitalist mode of production. In practice, as the so-called socialist societies insert themselves more fully into the market universe of world capitalism, contradictions similar to those of the capitalist mode of production and its technology tend to amplify. To paraphrase Marx yet once more, one cannot guess from the taste of wheat whether it has been grown under the watchful eye of the capitalist or under the picky control of the bureaucrat.

By elevating the ignorance of nature and life to new heights and adopting the mechanistic view of the world, derived from the Newtonian paradigm and promoted by apologetic versions of bourgeois political economy,[18] "official Marxism" has endorsed the most absurd choices in the field of development and therefore of resource utilization. Much as state ownership of the means of production fostered the "mentality of temporary workers"[19] among managers of the Soviet economy, the subordinate status of theory justified servile conformism among Soviet economists. In a society in which ownership and power are confiscated by the state, which plays the role of what Marx called "the abstract capitalist," private interests reappear under the perverse form of bureaucratic sectoral interests. The consequence is a productivism that has lost any human finality since it is not even subject to those brakes that in Western capitalist society emanate from antagonistic social interests. As the economist M. Lemechev writes:

With modern technology consuming large amounts of natural resources, production works more and more for itself. The example of iron ore alone is sufficient proof of this. In our country, 250 million tons . . . are extracted by the "progressive" method of strip-mining, thousands of hectares of fertile land are destroyed, and the hydrological cycle of vast regions is disturbed. . . . Then the steel industry causes pollution of the air and the water. The metal thus obtained is used to build giant steel rollers which manufacture sheet metal, which is used to build new giant steel rollers which manufacture sheet metal, which, in turn, is used to build new giant excavators for the mining of iron ore. The productivity of these excavators, the subject of pride for most engineers, is in reality a monstrous destructive force. A vicious cycle is thus created: a new technological cycle begins with disastrously minute results in terms of usefulness for human beings and tragically large results for nature.[20]

6. Toward an Eco-Marxism

The dismal ecological record of "existing socialism" is equally as disappointing as that of capitalism. A Marxism capable of understanding ecological economics and an ecologically sound socialist policy must take this truth into account. One must begin with Marx's famous remark in which he distinguished the part of the value of a natural resource that derives from human labor and the part that derives from "work" performed free by nature:

> So far therefore as labour is a creator of use-value, is useful labour, it is a necessary condition, independent of all forms of society, for the existence of the human race; it is an eternal nature-imposed necessity, without which there can be no material exchanges between man and Nature, and therefore no life. . . . We see, then, that labour is not the only source of material wealth, of use-values produced by labour. As William Petty puts it, labour is its father and the earth its mother.[21]

Marx focused his analysis on the capital–labor relation. The path opened by the concept of a society–nature totality, which could lead to a fruitful reflection on the interplay of social and natural determinations, was thus lost. This potentiality of Marx's Marxism remained unexplored. Marx himself contributed significantly to the extensions of his theoretical work in this direction by asserting—as early as the *Grundrisse* in 1857–1858—that capitalism's only limits were internal and that it was capable of emancipating itself from the determinants of nature.

Class struggle has been conceived only in its social dimension; material production has been strictly assimilated to an economic act. But every economic act is three-dimensional: economic, social–cultural, and natural. As has been noted, the latter aspect includes, on the one hand, all subtractions of energy and materials from resources, and, on the other hand, all waste rejected by the productive system into the environment. Yet this dimension remains almost absent from the Marxist schema of production and circulation. Marx emphasizes the natural origin of production on several occasions in *Capital*: "Labour is, in the first place, a process in which both man and Nature participate, and in which man of his own accord starts, regulates, and controls the material re-actions between himself and Nature. He opposes himself to Nature as one of her own forces, setting in motion arms, legs, head and hands, the natural forces of his body, in order to appropriate Nature's productions in a form adapted to his own wants."[22] Neither he nor his followers, however, developed this set of problems.

The keystone of Marx's analysis is the concept of value, already at the core of Ricardo's thought, and its sole source, labor: "Commodities, therefore, in which equal quantities of labour are embodied, or which can be produced in the same time, have the same value."[23] This definition attributes no intrinsic value to natural resources. In direct continuity with Ricardo, Marx emphasizes the point explicitly on several occasions: "It is thus strikingly clear that means of production never transfer more value to the product that they themselves lose during the labour-process by the destruction of their own use-value. . . . In this class are included all means of production supplied by Nature without human assistance, such as land, wind, water, metals in situ, and timber in virgin forests."[24] Humans therefore transform natural resources—wind, iron ore, trees, and the like— "into economic objects," but only the labor incorporated into these "objects" confers value (or more precisely, exchange value) to the latter. These are the premises of the Marxist analysis of the modes of appropriation of value created by labor; the difference between what the workers produce and what they consume is surplus value, the source of all forms of accumulation. But can a parallel not be established between this first mystification of the economy, the hidden mechanism by which surplus value is formed, and another, unsuspected by Marx, the hidden cost of things subtracted from ecological systems? Should the theoretical status of this concept of ecological cost not be ranked on a par with that of surplus value?

Marx wrote that "the elementary factors of the labour-process are 1) the personal activity of man, i.e., work itself, 2) the subject of that work, and 3) its instruments."[25] In the end, of these three poles of all production processes (nature, the human, and tools), Marx's analysis retains only the latter two. Nature is always hidden and its existence considered as primary, invariable data. Marxist thought has focused on human labor power as a commodity subject to the laws of the market, and on the producer as a subject of exploitation. The overexploitation of the natural environment and its appropriation by private individuals or the state, although often as important to the hegemony of classes or ruling social groups as the exploitation and appropriation of labor, did not receive equal attention. Only a few isolated Marxists, such as the Ukrainian socialist S. Podolinsky, attempted the difficult analysis of the role of the object of labor (nature) in production. Choosing the laws of thermodynamics as his tool, Podolinsky asked whether the labor theory of value was compatible with these laws: "According to the theory of production formulated by Marx and accepted by socialists, human labor, to use the language of physics, accumulates in its products a greater quantity of energy than had to be expended to produce the power of the workers. Why and how does such accumulation take place?"[26] The question is unsolvable according to a letter from Engels to Marx: "The energy value of a hammer, a screw or a needle, calculated according to the cost of production, is an impossible quantity. In my opinion, it is absolutely impossible to try and express economic relations in physical magnitudes."[27] Engels and Marx, uncritically engaged in the study of the concept of value as were all the economists whom they intended to criticize, completely missed the problem posed by Podolinsky. Yet these relations to the environment often cause irreparable damage to the majority of the living human population *and* to future generations. Much like the exploitation of labor power, they generate a discrepancy between immediate particular interests and the collective interests of society in the present *and* in the future. This gives rise to a triple contradiction:

- The overexploitation of a collective resource benefits the ruling classes, that is, private or state interests, rather than the interests of the majority.
- The distribution of costs and benefits over time is such as to bring profits to the present generation at the expense of future generations.
- The human species enhances its own interests to the detriment of those of thousands of other living species interacting within ecosystems in complex reciprocal ways.

7. Conclusion

Today the process of accumulation must be reconsidered explicitly in its global context, and resituated in the context of the resources and ecological relationships into which it has entered. One must therefore ask questions about the physical limits of all growth, and, more profoundly, about the entropic nature of all economic activity. Over the last 20 years many studies have demonstrated close links between systems of exploitation of nature and systems of exploitation of humans. Sharpening social afflictions such as poverty, rural depopulation, and unemployment are generally the reverse side of an unprecedented degradation of natural resources. The exhaustion of the latter *always* increases economic costs because the last natural productivity or resources must be compensated by energy substitutes and greater technological effort, not to mention the cost of repairing damage to the environment and human health. Lack of knowledge of the laws governing great biogeochemical cycles and the evolution of ecosystems can lead to irreversible breakdowns and jeopardize acceptable human life in the future. It is therefore also necessary to reformulate the transition to socialism in the direction of the restoration of a long-term sustainable balance between humanity and nature.

Considered rom the vantage point of historical ecology, the growth of industrial civilization has fed on a succession of disruptions of ancient local and regional ecological balances. Today the space of these disruptions has broadened to the entire planet, projecting human societies into a new relationship with nature: that of "world-ecology."[28]

As the century comes to an end the natural sciences have made it possible for the productive forces to achieve a power comparable to that of the geological and biological forces that have shaped our planet. The basic principle of capitalism is now orienting the economy toward the destruction of nature on a planetary scale. Socialist attempts to break this logic have failed since they presumed too many of its premises. The trend for the systems of production born of the Industrial Revolution points to a catastrophe scenario. The productivist economic ideology that justified them appears bankrupt. Whatever new solutions are explored by society, they will necessarily include a drastic revision of this productivist ideology.

We know the formula of the greatest economist of this century: "In the long term, we are all dead." What was only a joke has become the greatest challenge posed to the economy by the problems associated with ecology and bioeconomics. An immense job awaits us: after dissecting the mechanisms of the exploitation of labor power, we must now turn to those of the destruction of nature, of this nature that remains the material guarantee of the future of our species. Combining the two sets of problems remains one of the most urgent tasks. Do the words *economy* and *ecology* not have the same root, *oikos*, "the home," our home, the planet Earth?

Notes

1. I have tried to put this approach into practice in relation to energy; see J. C. Debeir, J. P. Deléage, and D. Hémery, *In the Servitude of Power—Energy and Civilization through the Ages* (London: Zed Books, 1991).

2. For a history of these relations, see M. Weber, "Energetische Kulturtheorien" (1909), in *Gesammelte Aufsätze zur Wissenschaftslehrer* (Tübingen: Mohr, 1922); and especially J. Martínez Alier, with K. Schluepmann, *Ecological Economics* (London: Basil Blackwell, 1987). This chapter owes much to R. Kaufmann's "Biophysical and Marxist Economics: Learning from Each Other," *Ecological Modelling, 38*, 1987.

3. David Ricardo, *On the Principles of Political Economy and Taxation*, in Piero Sraffa, ed., *Works and Correspondence* (Cambridge: Cambridge University Press, 1951), pp. 69–70.

4. Adam Smith, *The Wealth of Nations* (Oxford: Clarendon Press, 1976), vol. I, p. 37.

5. Karl Marx, *The Grundrisse*, ed. David McLellan (New York: Harper, 1971), p. 66 (from *Grundrisse*, pp. 73–77).

6. All the empirical data for this paragraph are taken from J. H. Ryther, "Photosynthesis on Fish Production in the Sea," *Science, 166*, 1969; P. F. Chapman, "Cout énergétique de la production primaire d'aluminium et de cuivre," *Cuivres, laitons, alliages, 130*, no. 3, 1974; C. Guillemin, "A propos de l'énergie . . . ," *Revue du Palais de la Découverte, 4*, no. 32, 1975; and C. A. S. Hall, C. J. Cleveland, and R. Kaufman, *Energy and Resource Quality: Ecology of the Economic Process* (New York: John Wiley, 1986).

7. Cited in Debeir, Deléage, and Hémery, *In the Servitude of Power*, op. cit., p. 146.

8. F. Ramade, "Crise de l'énergie, resources naturelles et production alimentaire," *Économie rurale, 124*, 1978.

9. R. Rosenberg, "Silent Spring in the Sea," *Ambio, 17*, no. 4, 1988.

10. Concerning the entropic dimension of the economic process, see Nicholas Georgescu-Roegen, *The Entropy Law and the Economic Process* (Cambridge, Mass.: Harvard University Press, 1971); Barry Commoner, *The Poverty of Power* (New York: Alfred A. Knopf, 1976); R. Passet, *L'Économique et le vivant* (Paris: Payot, 1979); and E. Tiezzi, *Tempi storici biologici* (Milan: Garzanti, 1984).

11. On this point, see Debeir, Deléage, and Hémery, *In the Servitude of Power*, op. cit., pp. 20–21.

12. Hall, Cleveland, and Kaufman, *Energy and Resource Quality*, op. cit.

13. C. J. Cleveland et al., "Energy and the U.S. Economy," *Science, 225*, 1984.

14. Concerning this point, see Debeir, Deléage, and Hémery, *In the Servitude of Power*, op. cit., p. 146 and note pp. 255–258, for a complete bibliography.

15. Karl Marx, *Capital: A Critique of Political Economy* (New York: Charles Kerr), book I, part 1, chap. 17, sec. 1, p. 417.

16. Ibid., book I, part 2, chap. 4, pp. 168–169.

17. M. Lemechev, "Sans dégrader la nature," *Les Nouvelles de Moscou,* 7, February 14, 1988, p. 8.

18. R. Passet, *L'Économique et le vivant,* op. cit.

19. B. Komarov, *Le Rouge et le vert* (Paris: Le Seuil, 1981).

20. M. Lemechev, "Sans dégrader la nature," op. cit.

21. Marx, *Capital,* op. cit., book I, part I, sec. 2, p. 50.

22. Ibid., book I, part II, chap. 7, sec. 1, pp. 197–198.

23. Ibid., book I, part I, sec. 1, p. 46.

24. Ibid., book I, part II, chap. 8, p. 227.

25. Ibid., book I, part II, chap. 7, sec. 1, p. 198.

26. S. Podolinsky, "Le Socialisme et l'unité des forces physiques," *La Revue socialiste,* June 20, 1880, p. 353.

27. Translated from the French edition of F. Engels, "Letter to Karl Marx of December 19, 1882"; reprinted in K. Marx and F. Engels, *Lettres sur les siences de la nature* (Paris: Éditions Sociales, 1973), pp. 109–110.

28. See my essay "From Ecological History to World-Ecology," in P. Brimblecombe and C. Pfister, eds., *The Silent Countdown* (Berlin/Heidelberg: Springer-Verlag, 1990), pp. 21–36.

Codependency and Indeterminacy: A Critique of the Theory of Production

Martin O'Connor

1. Introduction

The purpose of this essay is to retheorize the economic categories of production and technology in light of the current ecological crisis of capitalism and socialism. A fresh conceptualization is needed because both bourgeois and Marxist analyses of the ecological crisis are hampered by a concept of production and technology inadequate to present historical circumstances. In particular, an adequate theory of the "crisis in the conditions of production"[1] requires a reconceptualization of production itself. I will discuss two themes or lines of analysis: first, the problem of how to characterize production and technology in ecological perspectives, that is, the question of "nature" and its relation to the telos of "the development of the productive forces"; second, the "contradictions" inherent in the process of the capitalization of nature (i.e., of labor and the "physical environment") as a response by capitalism to economic crisis.

The issue of technology and productive forces has been a vexing one for Marxists who address the problem of ecological crisis. Two related factors have conspired to limit the scope and incisiveness of traditional Marxist analyses of ecological issues. On the one hand, it

Lecturer in Economics, University of Auckland, New Zealand. The author wishes to thank Juan Martínez Alier, James O'Connor, Bill Livant, John Foster, and John Ely for their help. First published in *CNS*, *1(3)*, no. 3, November 1989, pp. 33–57.

was often presumed that ecological destruction arises as a result of economic and social mechanisms specific to capitalist modes of production. On the other hand, traditional Marxist analyses of the contradictions of the development of capitalism and nonindustrial societies have embodied a Panglossian view of the growth of the productive forces.

The view of ecological destruction as a malaise specific to capitalism tended to go hand in hand with the supposition that the solution to ecological destruction will follow spontaneously from the establishment of socialist alternatives—an assertion that is theoretically suspect and empirically weak. Technology, in fact, is an abstract theoretical category that broadly designates the ensemble of a society's capacities for acting in the material world and for members of society to act on one another. The way in which we understand technology, as well as our conceptions of nature, are inflected by basic societal values; conversely, our societal values are inflected by society's technological prowess.[2] Subject to this dialectic, human beings themselves ascribe "value" and make value judgments, and hence give moral connotations to the known potentialities of the physical world and the people who live in it. The "ecological crisis" is thus a crisis arising from our social and moral conception of technology and its uses. To ascribe an innately beneficent *or* an innately malign potentiality to the forces of production in effect means that Marxism abandons any pretense of being critical. Nothing can be presumed a priori about the role of the "development of the productive forces" in the resolution *or* the aggravation of current ecological problems. Marxist theory that assumes that technology is "neutral" dovetails, in fact, with the presumption "that emancipation comes only via 'freeing' humanity from an (ontologically presumed) natural scarcity in which the possibility of social and political emancipation of humanity flows *with* the logic of accumulation."[3] This presumption must be excised for effective response to environmental crisis.

2. The Capitalization of Nature and the Naturalization of Capital

One leitmotif of ecological Marxism is the proposition that capitalism tends to impair or destroy its own social and environmental conditions and to retrieve itself from self-induced crisis through mechanisms and measures that cumulatively tend to worsen the damage.[4] Crisis induces capital to seek to exercise more control or planning over production conditions—labor, land, and space—as well as over the

production and circulation of capital itself. This process of the capitalization of nature signifies a deepening of political control over all
aspects of social life through a maturation and generalization of the
social institutions by which exchange value is instituted as a mode of
coding (signifying) and controlling social relations. It is an attempt
to treat all terrestrial life in the same way as produced commodities,
that is, controlled and subordinated to the finalities of capital.

The globalization of capital is, of course, somewhat spurious.
Yet it is an important ideological innovation. The capitalist system
undergoes a kind of mutation to its essential form, the culmination
of which would be the complete (notional) capitalization of nature in
which there no longer remains any domain external to capital. This
is tantamount to the assumption that an external nature does not
exist.[5] The image is no longer Marx's (or the classical economists')
of human beings acting on external nature to produce value. Rather,
the image is of the diverse elements of nature (including human nature) themselves codified as capital. Nature is capital, or, rather, nature
is conceived in the image of capital. The logic of the system is thus
the subsumption of all of the elements of nature-considered-as-capital
to the finality of capital's expanded reproduction.[6]

Theoretical difficulties immediately arise as a result of the fact
that this is a largely imaginary functional integration. The rhetoric
stresses harmonization and optimization; the reality is disorder and
conflict. As Baudrillard remarks, "Everything is potentially functional
and nothing is in fact."[7] Two sources of contradiction are inherent
in the process of the capitalization of nature, which furnish our justifications for proposing a shift from an industrial to an ecological Marxist perspective on production, on the "eventual" and "inevitable"
collapse of capitalism, and thence on the conditions for some sort of
socialism. The first is the fact that the planet is materially finite, a
situation that creates biophysical limits to the accumulation process.
The second, which is synergetic with the first, is the fact that capital
does not and cannot control the reproduction and modification of the
"natural" conditions of production in the same way it purports to
regulate industrial commodity production.

The first source of contradiction has been widely discussed in
the environmentalist literature. Many scholars and scientists have
pointed to the implications of thermodynamic irreversibility and the
fact that the earth is a materially closed system, hence unlikely to be
able to sustain recent economic growth trends.[8] These issues of natural
resource depletion, biophysical limits to growth, and so on, can easily
be housed within a Marxist perspective. In the past, the expanded
reproduction of capital has fed on the colonization of erstwhile exter-

nal domains—both social and physical. The globalization of the hegemony of capital obviously stymies this avenue of expansion (thus, for example, in the attempt to overcome this limit, we hear arguments for tapping extraterrestrial energy and mineral sources and for the use of deep space as a rubbish bin for waste).

The second source of contradiction pertains to the assumption, found in both bourgeois and traditional socialist "solutions" to environmental problems, that the "natural environment" is in fact a domain that can be controlled. The critique of this pretension requires an investigation of the "physical functioning" of economic systems. This essay develops a critique of orthodox views of production and the labor process from the standpoint of thermodynamics and open systems theory—a critique that describes the reality of the coevolution of a production system conceived as intimately codependent with an environment that is exterior to it and not controlled by this system (or, at least, imperfectly controlled by it). This critique presupposes a characterization of the complementary character of internal and external determinants of the functional organization and activity of a production system in thermodynamic terms, that is, as a system conceived as essentially open to its environment. Finally, it is necessary to work out the implications of this characterization of production for the theory of capitalism's environmental crisis as a "crisis of control." This line of reasoning will permit us to expose for critical scrutiny the presumptions underlying not only the image of the harmonization of production conditions within capital's designs, but also those inhering in the commodification of economic activity per se.

3. Theorizing Nature's Productions

A capitalist economy can be specified as an ensemble of processes of commodity production, that is, goods and services that are produced in processes directly controlled by capitalist economic agents, utilizing other commodities as inputs (the paradigm of an industrial or manufacturing process), the exchange of which is regulated through an integrated price regime (bureaucratic, marketlike, or mixed). If we wish to specify the conditions of production, we must be concerned with those factors that condition commodity production but appear to be external to it. All materials bearing on production and all bearers of "services" that are not themselves commodities may be classified as capitalism's conditions of production.

These definitions permit us to preserve the distinction, traditional in Marxism, between commodities produced in capitalistically controlled processes, on the one hand, and their social and physical conditions of production, on the other hand. Marx himself (as reconstructed by James O'Connor) conceived of three kinds of production conditions. The first was "external physical conditions," or the natural elements entering into constant and variable capital. The second was human beings themselves, or labor power, the "personal conditions of production." The third was "the communal, general conditions of social production," or the social and physical infrastructure that shapes, and is shaped by, human action. Capitalism as a process tends to effect the ideological production of "capitalist nature," where this term denotes "everything which is *not* produced as a commodity but which is treated *as if it is a commodity*."[9] The distinction between commodities in the narrow sense, and "capitalist nature" in the sense of elements that are treated as if they were commodities, thus refers to the difference between real and direct control, on the one hand (as in a manufacturing process), and a formal or nominal control, on the other hand. In the latter case, the element of nature (or human nature) is *imaged as if* it is produced and reproduced through the controlled allocation of inputs by economic agents. However, real and direct control is not maintained.[10] Our focus thus needs to be on the contradictions related to the nonattainment of real control over external (and human) nature.

Marx himself wrote relatively little about the ways that capital impairs its own social and environmental conditions, although he was aware of the environmental and human degradation associated with industrial processes. In particular, he never theorized the *feedback* dimension of the interdependencies between capitalist production and its natural cum social environment. It has been suggested that this lacuna was due in part to the fact that "historical nature [in Marx's time] was not capitalized to the degree that it is today."[11] While this is true, and while the process of nominal capitalization has today attained unparalleled proportions, some key reasons lie elsewhere.

Marx held a view of the "production process" that, typical of his time, gave an impoverished and partisan picture of the physical interdependency of a productive system with its environment. In the famous description of the labor process in part 3 of *Capital I*,[12] Marx describes labor as "a process going on between man and nature, a process in which man, through his own activity, initiates, regulates, and controls the material reactions between himself and nature." Marx thus adopts a tacit premise of controllability. Man (*sic*) "confronts nature as one of her own forces, setting in motion arms and legs, head and hands, in order to appropriate nature's productions in a form suitable to his own wants." The finality of labor is determinate:

> The labour process ends in the creation of something which, when the process began, already existed in the worker's imagination, already existed in an ideal form. . . . In the labour process, therefore, man's activity, with the help of the instruments of labour, brings about changes in the subject matter of labour, changes intentionally effected. The process disappears in the product. The product is a use-value, materially supplied by nature, and adapted to human wants by a change of form.

The labor process is thus "purposive activity carried on for the production of use-values, for the fitting of natural substances to human wants." In few passages in Marx is there any hint of unintended and undesirable by-products and unavoidable side effects of productive activity. What is the use value of the hole in the ozone layer? What wants are satisfied by the production of nuclear wastes? What is the final "product" in the process of atmospheric and biological change being unleashed by modern production technology? The experience of more than 100 years of development in the instruments of labor leaves the rather stark message that the capacity to substantially intervene in a social or natural system is very far removed from a basis for determinate regulation of the forces of nature.

Marx was, of course, conditioned by his time and place. Subsequently, most Marxists have owed more than they have realized to the view of nature as brutish and inert, as well as to the fascination with progress and control, which was manifest in post-Enlightenment scientism in general and in 19th-century industrialism in particular. It is important to note this intellectual dependency today, and its lacunae, as a precondition of efforts to go beyond it. In fact, the theorization of nature (and hence of production) is an aspect of praxis, and thus is historically open, never definitely achieved. It must be conceived as a historically specific endeavor, not something that can be accomplished once and for all. Indeed, it is arguably never accomplished.[13]

To highlight the blind spots of Marx's conception of production and of the labor process, I will draw on elements of contemporary "open systems analysis" and the thermodynamic theory of irreversible processes. No pretense is made of standing above these dialectics of nature's theorization. Thermodynamic science is a child of a particular time and place, spawned in the heart of the 19th-century industrialization process of Western Europe. The "discovery" of energy was not an autonomous theoretical advance that occurred independently of social and cultural conditions, and that simply was put to subsequent service in social or engineering applications. The cultural milieu of the times both motivated the emergence of thermodynamic theory and also powerfully and fundamentally conditioned the interpretation it received at birth and in its applications ever since.[14]

Put another way, prevailing images of nature reflect the relevant central motivations and orientations of a society. The interpretation of capital as embodying "stored" labor power, and of nature as holding potentials able to be unleashed to augment the productivity of human labor, was a central social reality of 19th-century industrializing societies. Wolfgang Sachs writes that

> from John Locke via Adam Smith to Karl Marx, man has been increasingly perceived as possessing labour power, that extraordinary capacity which allowed him, in the eyes of these economic thinkers, to add new wealth to the world. Work was considered to be a *productive* force. Endowed with this capacity, man no longer had to confine himself to the skillful appropriation of the given riches of the world; rather, he could accumulate additional wealth and create infinite progress.[15]

The 19th-century view of energy intimately reflected this underlying social preoccupation with the capacity of men and machines to "do useful work." Prigogine and Stengers termed this worldview "a conception of society and men as energy transforming engines."[16] On the one hand, the conception of man as "worker" (as in Marx's own analysis) served as an image or metaphor through which to comprehend the machine, and by extension, nature as amenable to transformation through labor and machine. On the other hand, the scientific concept of energy, itself emerging out of this historically specific image of nature, contributed in turn to the amplification of this image. As Sachs observes,

> The steam engine was capable of performing work, thereby surpassing and replacing human work. Heat, now convertible into movement, could be measured by its capacity to "work," that is, to lift a given weight to a certain height. Why not interpret other forces of nature with the same model? Once nature itself had been perceived as a working machine, all its forces could be compared and evaluated by their ability to perform work.[17]

The concept of and construction given to "energy" thus did not fall from the sky like Newton's apple. Rather, to speak of energy was specifically to evoke the immense powers of nature potentially at work for progress in industrial production. If thermodynamics furnished insights into the nature of production, it is equally true that prevailing notions and conditions of "production" inflect profoundly the particular senses and social weights that have been accorded to the thermodynamics results. Any recourse to thermodynamics, therefore, is open to misadventure, as a result of the symptomatic myopia it may induce. The fact that thermodynamics is a product of industrial society lends it a double-edged pertinence: first, as a tool of ideology, and second (as I will try to employ it) as a tool of immanent critique.

Concern for energetic constraints on economic activity, for example, has recurred frequently since the mid-19th century.[18] The glowing representation of nature's forces, of course, had its dark side, namely, the phenomenon of energy dissipation. Not only is the yield of useful work from any energy source absolutely constrained, but all productive processes result in a "using up" of energy available to do work, meaning there is a "net loss" of available energy for future use. This result, codified in thermodynamic terms as the irreversibility of entropy production, gave rise to the specter of "heat death." Forms of energy that could be tapped to fuel economic progress might be limited, and could be exhausted in the future. If classical thermodynamics imaged nature and society based on the model of the steam engine, then nature was "a reservoir of energy that is always threatened with exhaustion."[19] The second law of thermodynamics, which postulates entropic irreversibility, in this way added its distinctive weight to the imperative of efficiency in economic production.

There is thus nothing new or "radical" in the modern preoccupation with natural resource scarcity. Moreover, the recognition of a "resource depletion" problem has frequently been the occasion for simple reiteration of the technoeconomic "fix" of better control and improved efficiency of natural resource use. The task I set myself here is quite different from these technocratic concerns. It is to use the insights and results of thermodynamic theory to undermine the pretended "scientific" foundations of the discourse of control common to both capitalism and mainstream socialism, that is, to highlight the "coevolution" of conditions with conceptions of production and of nature (and human nature) with the ways in which they are theorized.

It is important to stress the idea that the environment of a system is delineated dialectically, referring to whatever is "other" to the system of analytical interest. This boundary signifies a relationship or, better, engenders a relationship. The analytical relevance of environment rests on the recognition that often the behavior of a system can be comprehended only through considering that system's interactions and exchanges with its environment. This notion of interdependence is the kernel of the concept of an *open system.*[20]

In practice, there exists a hierarchical structuring among the component parts of reality. Hierarchy means (following Herbert Simon) a set of Chinese boxes where each successive layer of disaggregation discloses many somewhat comparable boxes, for example, cells in a living organism. In thermodynamic perspective one can consider the global (terrestrial) system as such a hierarchical array of subsystems. The global *ecosystem*, along with each constituent subsystem (whether economic or geological–ecological) can be thought of as a single "production process," or as an ensemble of smaller processes. These may be processes of no apparent finality (such as geophysical processes), or, alternatively, processes may be regulated from external sources so as to form a controlled unit within an organized greater whole (such as metabolic or industrial production processes). Or they may be essentially self-regulating (as in "self-organizing" or "autonomous" biological and social systems—the "market" being a putative prime example).

Such processes, whatever the level of aggregation, will be mutually interdependent, and need to be conceived of and analyzed as systems that are open in two distinct but connected senses. First, they are open to their environment in the physical (thermodynamic) sense that each will receive from and provide to other processes inputs and outputs, respectively, of material resources, and there will be mutual exchanges of energetic services (heat and work exchanges); second, they are also open in an informational sense, having the potentiality for a "becoming" through time, manifesting an irreversible, unpredictable, and more-or-less unique and novel evolutionary behavior emerging from the interplay of internal and environmental change factors.

In short, we are concerned with the coextensive, codependent activity of an ensemble of more-or-less mutually coordinated production processes, both economic and "geological–ecological." In particular cases, a system may be said to be thermodynamically dominant in relation to its cosystem (environment), or vice versa, in one of two senses.

The first may be called *strong thermodynamic dominance*, where there is maintenance of complete control over the organization and dynamic behavior of its cosystem. This is not the norm in the world. It is, however, the norm in casual "scientific" discourse. Much popular wisdom about scientific method relies on conceptualizing systems in this way, for example, experimentation based on a "control" or "reference" system relative to which the impacts of selected changes in "inputs" are tested. Also, this is the conceptualization that underlies the economist's traditional conception of a production process.

The second may be denoted *weak thermodynamic dominance*, the case where the productive activity of the cosystem is not completely controlled by the system, but where system-environment interactions are regulated so that the latter has insignificant perturbing effects on the structural organization and evolution of the system. One can imagine a system that successfully "exploits" its environment without changes in environmentally organization having any significant "feedback" effects on system activity.[21]

From a theoretical point of view, the crucial point is that situations of thermodynamic dominance, whether weak or strong, are special cases. The behavior of most systems is not totally determined by environmental boundary conditions, yet is nonetheless vulnerable to perturbation from the outside. In other words, the internal activity of each system is to some degree autonomous with respect to the inputs, extractions, and boundary conditions "imposed" on it by its environment, yet remains conditioned by these later.

If a productive system is considered from the standpoint of its interdependency with an exterior domain, the keynote is openness and uncontrollability, hence instability. Apparently insignificant "small" changes in environmental conditions can radically alter the modes of activity displayed by the system, and hence its subsequent interactions with its surroundings. Moreover, whatever the order of system considered, and whatever the propositions made about its internal autonomy or finalities, there will be always be an exterior domain that is not amenable to full control and hence that may bring about the undoing of the system's designs. This is the role of environment as a system's Nemesis.

If we consider a system specifically from the point of view of its organizational autonomy, the keynotes are a latent creativity and instability; that is, there will always be a degree of indeterminacy associated with the behavior of the elements of any system. We can refer to this property as the "complexity" of the system.[22]

Yet neither natural nor social history can be written treating their distinctive object material as set in watertight phenomenal domains where separate disciplines can (each in their own way) aim to arrive at a "celestial mechanics" of their subject matters. In formal systems' terms, the stimulus to change works reciprocally between many different levels. Changes in "macroscopic" properties and behavioral modes of the system under analysis may be induced by component subsystems, as a result of the "microscopic" diversity that underlies the emergent whole. Conversely, changes in whole-system behavior may facilitate or induce transformations in subsystem behavioral modes. Moreover, since every system is a subsystem from another point of view, changes in external (environmental) conditions may induce, through violence or through internal amplification of small perturbations, qualitative behavioral changes within the system itself. The respective histories of any two (or more) systems will each be characterized by cascades of instabilities and catastrophes, in each case deriving from the interplay of respective internal and external influences.

4. Unmasking the Industrial Production *Épistémé*

It is now possible to consider critically the conventional way of specifying a "technique of production" in the main body of economic literature. A raft of controllability assumptions underpin the entire corpus of the modern theory of production and growth.

1. *Controllability of a production process.* Usually a technique, or a spectrum of feasible techniques, is represented through relating specified ratios of inputs of economic resources (i.e., inputs of materials and energetic services directly controlled by economic agents) to specified levels of outputs of economic resources. For such a specification, complete control is presumed over the combination and reaction of the inputs. This amounts, in effect, to the assumptions of strong thermodynamic dominance by production managers over each process of commodity production. The inputs are transformed according to determine rules or know-how, so that the process can be conceived as leading to determinate output results. Perfect functionality is the norm. Unpredictability is attributed to error or "accident."

2. *Dominance over the environment.* Further, it is always implicit that complete control can be and is maintained in relation to all non-economic (i.e., environmental) processes, insofar as these latter are interactive with economic processes. Traditionally, in economic analysis, the physical (or "natural") environment has been considered as providing essentially unchanging conditions as a site for economic activity. At or around a given site, "natural resources" can be freely appropriated up to the limits of their availability. For example, land was considered by the classical political economists as fixed in total quantity, at least as a first approximation. Amenities such as air, sunshine, scenery, and so on, were (when they were discussed at all) construed as available "for free" within given limits for the use or enjoyment of members of society. Economists call this the "free gift" assumption applied to the environment. Similarly, economists have traditionally assumed "free disposal"; that is, they have supposed that wastes and excess outputs from economic production processes can be assimilated by the environment without affecting the economic production processes themselves. In sum, the environment makes its

appearance, if at all, merely as a set of constants (constraints and invariant parameters) or blank slates (sinks, etc.). This amounts, in effect, to an assumption of weak thermodynamic dominance of an economic system over its natural environment (i.e., no significant uncontrolled feedback implications from environment to economy). It is what Charles Perrings calls the "weak environmental assumption" in economics: "that an environment exists; that it is not completely dominated by the economy, but that it plays only a benign and passive role."[23]

3. *Independence of production processes from each other.* It is further assumed that, as a general rule, the technique of each individual process can be expressed without explicit reference to the levels of activity of other economic processes within the system. It should, in this regard, be noted that the complement of processes to any chosen production processes within an economic system will as a matter of definition make up a part of the latter's physical environment. Repression of all reference to process–environment interdependencies is possible only on the tacit assumption that environmental effects on production process activity are known and/or constant for any particular context. Given the dialectical symmetry of the situation, this means assuming reciprocal weak thermodynamic dominance of each process over all others, that is, that they do not perturb each other, or, at the limit, that they are wholly independent of each other.

The basic picture of "production" is thus that engineers/managers regulate the inputs and environmental conditions with a view to obtaining a predetermined output from a "black box" process. Paradigms would be the steam engine and the factory assembly line. This ensemble of presumed control properties can be called the "industrial production *épistémé*" (IPE). The term *épistémé* (used in the sense of Michel Foucault) connotes a particular manner of knowing reality. Taken together, the whole ensemble of control assumptions amounts to a particular mode of conceptualizing production activities. The adjective "industrial" signals that this manner of representation of production has emerged into prominence in the context of the development during the past 200 years of Western "industrialized" societies. Other societies have understood production and technology, or the action of transformation of material reality, in quite different ways.

Traditional Marxism and bourgeois economics have shared much in common in their tacit presumption of the essential controllability of nature through the agency of technology. The effect of the industrial *épistémé* assumptions is systematically to negate, or to "bracket out" under *ceteris paribus* clauses, the mutual conditioning and interdependency of production processes. Taken together, these assumptions allow any commodity production process considered singly, and hence the economic system taken as a whole, to be represented as a closed information system in the sense defined by Anthony Wilden, one that "in reality or by definition, is not in an essential relation of feedback to environment."[24]

This is, of course, exactly what is not true. Current awareness of the interference afforded by juxtaposed production activities, of the significance of infrastructures, and of the pollution, ecological derangement, species extinction, and material resource exhaustion consequences of economic activity, makes this only too clear. However, this recognition of the significance of economic–geological–ecological interdependencies has not led to a widespread interrogation by economists of their controllability assumptions. Rather, as mainstream economic theory has enlarged its scope to embrace (pro forma, at least) the phenomena of environmental crisis, the tendency has been to extend the same sorts of assumptions to the erstwhile social and ecological conditions of commodity production. This leads to what Perrings has called the "strong environmental assumption," which "supposes that the economy completely dominates its environment."[25]

This formal analytical extension of control assumptions closely parallels the enlargement of the domain of the political hegemony of capital. An interrogation of the contradictions and impasses of this theoretical extension thus provides us with a paradigm useful for a critical understanding of capital's "desire" for control, and of the social and political processes, and their contradictions, fueled by these imperatives. These two developments, one overtly political, the other putatively scientific, are intimately related. Dominant discourses in and about science inform the general drift of economic theory, which informs in turn the political process. Prevailing political/ideological exigencies influence scientific ideology in general, and encourage (and overtly reward) the elaboration of discourses of management and control that suit the order of the day.

The extension in this way of the scope of the industrial *épistémé* assumptions to the natural and social conditions of commodity production, is, however, a largely imaginary process. Capital does not and cannot actually control all aspects of the reproduction and modifications of the conditions of production. Inasmuch as human societies, and more widely the biophysical milieu of commodity production, cannot in reality be wholly dominated by capital, or by economic agents acting consistently with or subordinated to the imperatives of capital, perturbation and feedback effects on and within the production system must remain both real and indeterminate. The attempt to salvage the industrial production *épistémé* through its generalizations simply avoids confronting the essential problem of the environmental crisis, namely, the openness of any system of production to unanticipated and uncontrolled perturbations arising from the environment that conditions it.

5. Toward an Ecological Production *Épistémé*

The time is overripe for a repudiation of the industrial *épistémé* assumptions. It is true that some production processes are controllable in accordance with the engineering model of determinate input–output relations. In general, however, in the case of complex systems, this is not so. The message of modern physical and life sciences is not unambiguously that more knowledge about nature's forces means better control; rather, science raises new and sharper questions about the indeterminacies attaching to our interventions and attempts at control.

Indeterminacy of the time path of a complex system—that is, concerning the "outputs" that will obtain from some given initial state, inputs, and environmental conditions—implies the impossibility of predicting in advance, and perhaps in any way controlling, the "inputs" that this system will provide to others codependent with it. Autonomy and openness are inherently linked to indeterminacy and incomplete control. The open systems view of reality thus yields a picture of a turbulent coevolution of interacting "entities." These latter are persisting structural forms arranged in hierarchies that are, however, interlaced, confounded, and concatenated.

Such a conceptualization of the codependency of autonomous production activities may be termed an "ecological production *épistémé*" (EPE). It is radically opposed to the industrial *épistémé* in several respects. The IPE conceives of individual production processes as essentially separable, and also controllable both individually and in the aggregate. For the industrial *épistémé*, time connotes continuity and predictability, more specifically, an irreversible accumulation: the expanded reproduction of capital. The ecological *épistémé* also insists on time irreversibility. But here the emphasis is on time as the axis of change, and on the incompleteness of systemic evolutions. Time is, in fact, the Nemesis of any pretension to have determined the future, and connotes the indeterminacy, ambivalence, and mutability of historical trajectories.

It will often be possible to specify some "underlying" laws that remain valid even while identities or regularities are violated (genesis, catastrophe, etc.) at "higher" levels. However, these laws will only constrain, and generally not determine, the possible and actual forms taken by the higher level activities, and are themselves mutable under sufficiently "extreme" conditions. Over time, one will observe the coming into being, maintenance (reproduction), or disappearance (breakdown) of discernible entities. Over the relevant time scales, this is as true for any regime of expanded reproduction of capital as for a biological organism. Both are as evanescent as smoke rings in the wild rhapsody of history. In general, observable regularities are trumped by the possible upending or collapse of any and all observable laws and phenomenal regularities. The concatenation of locally definable identities and regularities is not, in general, resolvable in terms of a "global" identity. Whatever might be the hegemony of equilibrium concepts in ecology as well as in economics (and this is a matter of scholarly dispute), there is no "whole of being" definable as a unity or identity that determines or contains the incessant interplay of different orders of change and constraint.[26]

6. The Capitalist Crisis of Control

The final purpose of this essay is to highlight certain aspects of capital's crisis of control through a heuristic use of concepts from thermodynamic "open systems" analysis. This is in no way a reductionist analysis. Thermodynamic concepts and measures have no special privileged status in the view of coevolution sketched above. Rather, it is axiomatic that forms of description must be sought that are appropriate to the particular phenomenal domains in question. This question of pertinence must always be decided locally, that is, as a function of the particular time and place, and of the circumstances and purposes of analysis.

Nowhere is this more obvious than with social processes, where the mutation of historical forms radically conditions the meaningfulness of particular analytical categories. Modes of analysis that seem pertinent in one instance—that is, that render intelligible a particular course of events in terms of various regularities, phenomenal laws, necessary relations, and so forth—may in the face of change become defunct or give way to other modes of analysis. Thus it should be clear that in proposing an "ecological production *épistémé*" I am in no way stepping outside of history. My appeal is for the pertinence of the conceptual framework for us here and now, rooted as we are in a particular cultural and material history.[27]

On one level, the specific epistemological emphasis of the EPE is that, in our role as analysts, we are not observers of society, but observers–actors in society. On an analytical level, the specific emphasis of the EPE (in contradistinction to the IPE) is that we do not act *on* nature so much as act and interact *in* nature. The key issue is to specify the qualitatively most important features of this action-in-nature. This must be done not in terms of some abstract essentialism of the labor process, but rather in the sense of defining particular properties and "moments" that allow us to grasp the phenomena of capital's ecological crisis and to render it intelligible. It is in this sense that I have highlighted lacunae in Marx's view of the labor process, and have insisted on a common ground of capitalist social and ecological crises in the noncontrollability by capital of "external nature" (both in animate and inanimate) and human nature.

In actuality, terrestrial nature is only incompletely capitalized, and that more often than not only notionally. All prospects for stability in a system of commodity production, or for a harmonious "expanded reproduction" of a system of commodity-capital or capitalized nature, will founder somewhere between the Scylla of noncontrollability due to elemental volatility from within, and the Charybdis of openness to uncontrolled exterior domains. The absence of full control of exterior domains, social as well as natural, should be self-evident. For example, the habits and relations of production of people in "traditional" societies, and the institutional arrangements of these societies, are frequently encountered as "obstacles" to the penetration of capital, and have widely been theorized as such. But these "hindrances" are not a matter of the mere "inertia" of these nonindustrial societies but rather evidence of their own dynamism according to modalities and imperatives different from (and often antagonistic to) capital's own.[28]

In ecological Marxism, therefore, it is the *reciprocal* character of the interplay of a system of capitalistic production and its "natural" conditions that must be highlighted. Attention needs to be focused not only on the way in which capital acts on its external conditions, that is, on the dynamics of capital's transformation and attempted control of conditions, but also on these conditions' "resistance" to, suborning of, and subversion of this attempted control. This resistance may be construed along two lines; first, as involving a subjective or self-conscious dimension (notably human), and second, as the de facto resistance of reality to its theorization as controllable in capital's imaginary. The fact that the "objective" noncontrollability of nature concerns us, and the very existence of self-conscious social and political movements, is proof of the "subjective" dimension. Attention to both is necessary for any theory and practice with genuine critical aspirations.

As much as it neglects or downplays the implications of physical openness to an exterior environment, the functionalist metaphysics of industrial and postindustrial capitalism also negates the irruptive potentiality of every element of nature. The particular emphasis of this essay has been on the contradictions inhering in the notional internalization by capital of erstwhile external and uncontrolled domains. Under the presumptive extension of industrial *épistémé* control assumptions to erstwhile "external" and "human" nature, these latter are analytically dissociated into an ensemble of component objects amenable to functional integration, similar to the factor inputs of a manufacturing process. But the real effect of idealizing nature as controllable is to "produce" a counterphenomenon, the elemental volatility of nature imaged as capital. If for industrial production the paradigm of dysfunction was the accident, today, concomitant with the capitalization of external nature, the instances and risks of "accidents" seem to have become global in their ramification. Nuclear reactor malfunctions and runaway genetic experiments are paradigms of the "ecological accident," or rather of the accident ecologized. And it is no accident that the specter of accidents so haunts the modern consciousness. A kind of metaphysical principle of "the Accident" is simply the dialectical by-product of the ideology of mastery of nature by technology.

The accident is thus a sign of the nonrealization and nonrealizability of the capitalist (and traditional socialist) ideal of perfect functionality. The phenomena of riot, catastrophe, and accident, in their peculiar "senselessness," are, in effect, cogenerated alongside the deepening of the project of rationalization of nature through the technology that underpins modern capitalist and socialist (state capitalist?) social systems. It is as if, relative to capital's finalities, and its agents' technocratic designs, "a malign demon is lurking about to ensure that the beautiful machine always goes off the rails."[29]

There appears, in effect, a refractoriness at the heart of every part of capitalist nature. Everything happens as if these elements "resist"—in arbitrary, whimsical, and "irrational" fashions—their attempted codification within capital's designs. We could say that they display an incurable volatility. They do not necessarily do what is "needed" for the system's stability or "equilibrium," yet will do any variety of other things. The elements of "capitalist nature" display, from the standpoint of their possible subordination by economic agents to capital's finalities, a lamentable intractability. Any functional harmonization that may be construed to exist between an element of "nature" (whether capitalized or not) and the "economic environment" around it will be menaced incessantly by the irruption of novel and incongruent behavior in that element.

Again, this is apodictic. Whether considered in its thermodynamic dimensions as I have done here, or in its social and political dimensions (unruly masses; diverse manifestations of "irrationality" and revolt by people against capital in the name of autonomy, difference; and so on), the refractoriness of nature in the face of its attempted determination by capital (or by any other single finality) is, if we only look, on show everywhere. It is evident in the seemingly gratuitous variety and changeability of natural phenomena, and equally in the inventiveness (if not outright perversity) of human nature in all its diverse sociocultural modalities.

7. Conclusion

The generic question I have posed is the following: What specific forms of contradiction arise in consequence of the fact that social and geological–ecological production activities are refractory to capital's project of their determination? This cannot be answered a priori. Not only is there considerable diversity, even within Marxist circles, concerning the appropriate mode for analysis of capital's "laws of motion," there is an even more radical diversity in the particular, local "resistances" that capital encounters. The permutations further multiply once it is admitted that capital's ways of accommodating itself to these symptomatic crises, large and small, will depend to some extent on circumstances and types of resistance. For example, the formal (nominal) subsumption of non-Western societies and non-industrialized minorities within the hegemony of global capital (valorization of the "informal sector," ethnocommodities, etc.) is accompanied, in some Third World societies, at least, by a real disintegration of economic and wider cultural fabrics, both "traditional" and "modern." The outcome of the interplay of these two tendencies is anything but certain.[30]

The insistence in the EPE on the reciprocal character of action, and on the "resistance" and refractoriness of nature, is an important innovation for ecological Marxism by comparison with traditional Marxism. Although usually resistance to capital has been posited (as spontaneous or otherwise) on the part of oppressed classes, the Marxist view of the interaction of human with nonhuman nature has been curiously one-sided. Marx's own view of the labor process does little to dispel the IPE ideology of control and of a determinate course of human history. Labor "makes use of the mechanical, physical, and chemical properties of things as means of exerting power over other things, and in order to make these other things subservient to his aims."[31] But subservience is hardly nature's way. Rather, as a sort of antithesis to the Enlightenment/industrial project of control (to which Marx was, in the final analysis, as much party as critic), everything happens as if nature "resists" this (imaginary) codification whether as in-the-service-of-capital or in-the-service-of-labor.

In fact, relative to any attempted determination at all, the elements of nature, human and otherwise, nominally capitalized or not, hold within them a sheer excess of potentiality, a latency of multiple possibility, a capriciousness, an illegible ambivalance.[32] All of this suggests that the telos of capitalist crisis will have to be deemed multiple and indeterminate. Along the way, the labor process itself needs to be respecified in "ecological" perspective. To this end, the following questions may be raised: How might one theorize an "ecology of action" departing from the admission of an irreducible autonomy of each element simultaneous with its being in intimate interdependence with its surroundings? Or, in more Marxian terms, how might one theorize the "unity" of social production, at the same time according weight to the specificity of each "site" of productive activity? How, furthermore, might one accommodate the obvious purposefulness of human activity, to the constant possibility of its reversal or suborning by a countervailing activity, human or otherwise?

The seeds of some possible responses to these questions have already been planted. The peculiar cogency of "site-specificity" to the present debate[33] arises not just from the legitimate resistance to the homogenization process imposed by abstract capital. It also is "produced objectively" as the antithesis of capital's atomizing metaphysics. The latter's pretension of instrumental control and functional integration is predicated on a view of nature (and human nature) as an ensemble of separable objects, each with distinct and determinable properties and each ideally substitutable and replaceable. The ecological *épistémé* suggests a radically different view. Although productive action, human and otherwise, may be localized in its origin and intent, the sense and effect of all human action depends radically on its social as well as its material context. Every action is, therefore, both irreversible and indeterminate in its "global"significance, and, in this sense, every event is unique and unreproducible. The EPE picture of a turbulent coevolution locates the essential "production" at the level of the whole "ecosystem" of interest. It is this whole that, through the synergy of its coextensive parts, engenders its own re-production, re-sourcing, re-creation, and re-novation. The problematic of site-specificity requires us to address both the irreducible difference and radical codependency of events, and of the sense(s) that they may take in history. It requires us to address and give theoretical and existential standing to the phenomena of violence (unilateral negation of another's intent) and of seduction (being taken in the design of another). All this is masked over and obfuscated by the timeless "value form" of abstract capital. It is nonetheless perfectly visible in the manifold resistances that flower in the cracks and interstices of the dysfunctioning world economy.

Notes

1. James O'Connor, "Capitalism, Nature, Socialism: A Theoretical Introduction," *CNS*, *1(1)*, no. 1, Fall 1988.

2. See, for example, Richard Lichtman, "The Production of Human Nature by Means of Human Nature," *CNS*, *1(4)*, no. 4, June 1990, pp. 13–51, and R. Norgaard, "Sustainable Development: A Coevolutionary View," *Futures*, *20*, December 1988.

3. As phrased by John Ely, "Lukacs' Construction of Nature," *CNS*, *1(1)*, no. 1, Fall 1988, p. 111.

4. James O'Connor, "Capitalism, Nature, Socialism," op. cit., passim.

5. Charles Perrings, *Economy and Environment: A Theoretical Essay on the Interdependence of Economic and Environmental Systems* (Cambridge: Cambridge University Press, 1987), pp. 4–5.

6. The sense of this mutation is evoked by Jean Baudrillard, "Design and Environment, or How Political Ecomony Escalates into Cyberblitz," in *For a Critique of the Political Economy of the Sign* (St. Louis: Telos Press, 1981), English translation by Charles Levin of *Pour une critique de l'économie politique du signe* (Paris: Gallimard, 1972). For an expanded discussion, see my "On the Misadventures of Capitalist Nature," Chapter 7, this volume.

7. Ibid., p. 197.

8. Some illustrative references in the now abundant literature include: G. A. Daneke, ed., *Energy, Economics, and the Environment* (Toronto: Lexington Books, 1982); Nicholas Georgescu-Roegen, *The Entropy Law and the Economic Process* (Cambridge, Mass.: Harvard University Press, 1971), and *Energy and Economic Myths* (New York: Pergamon Press, 1976); Malcolm Slesser, *Energy in the Economy* (London: Macmillan, 1978); René Passet, *L'Économie et le vivant* (Paris: Payot, 1979); J. C. Debeir, J. P. Deléage, and D. Hémery, *Les Servitudes de la puissance: Une Histoire de l'énergie* (Paris: Flammarion, 1986).

9. James O'Connor, "Capitalism, Nature, Socialism," op. cit., p. 7.

10. As Bill Livant has pointed out (pers. comm.), there is some sort of parallel here with the distinction made by Marx, in a different context, between the "formal" and "actual" subjection of labor to capital, the transi-

tion from the former to the latter being concomitant with imposition of the capitalist mode of production.

11. James O'Connor, "Capitalism, Nature, Socialism," op. cit., p. 15.

12. Karl Mary, *Capital* (London: Dent, 1930), chap. 5. pp. 169–177.

13. See Lichtman, "Production of Human Nature," op. cit.; Cornélius Castoriadis, *L'Institution imaginaire de la société* (Paris: Seuil, 1975), English translation by Katherine Blamey: *The Imaginary Institution of Society* (Cambridge, Mass.: Polity Press, 1987); and Serge Latouche, *Le Procès de la science sociale* (Paris: Anthropos, 1984).

14. Wolfgang Sachs, "The Social Construction of Energy: A Chapter in the History of Scarcity," Working Paper, Technische Universität, Berlin, September 1983; Klaus Schlüpman, "Natural Resources, Productive Forces, Semiotization of Science: The Case of Rudolph Clausius," paper presented at to the 2nd Vienna Centre Conference on Ecology and Economics, Barcelona, September 1987.

15. Ibid. He adds: "In Voltiare's time, the clock served as a metaphor to interpret nature in a coherent and meaningful manner. In that image, nature appeared as a complicated but well-ordered clockwork, incessantly in motion, wisely and purposefully arranged by God, the original clockmaker." The transition from clockwork to steam engine is thus congruent with the gradual shift from mercantilist and physiocratic concerns with land and precious minerals as fixed sources of wealth, to the labor theory of value of the classical political economists.

16. Ilya Prigogine and Isabelle Stengers, *Order out of Chaos* (London: Heinemann, 1984), p. 111.

17. Sachs, "The Social Construction of Energy," op. cit.

18. W. S. Jevons writing around 1865 on "The Coal Question" in Britain is one example. Juan Martínez Alier, in *Ecological Economics* (London: Basil Blackwell, 1987), gives a comprehensive historical survey.

19. Prigogine and Stengers, *Order out of Chaos*, op. cit., p. 111.

20. A more detailed discussion of the thermodynamic description of codependent systems and issues of controllability in ecological economics is provided in my "Physical Functioning and Economic Control: Thermodynamic Arguments for a Co-evolutionary Perspective in Environmental Economics," Working Paper no. 29, Department of Economics, University of Auckland, 1986; see also my *Time and Environment* (Ph.D. diss., Department of Economics, University of Auckland, 1990); and Oliver Godard, "Autonomie socio-économique et externalisation de l'environnement: La théorie néoclassique mise en perspective," *Économie Appliquée, 37,* no. 2 (1984), pp. 315–345. A comparable exposition of the view of economic production processes as embedded in a larger ecosystem can be found in Perrings, *Economy and Environment*, op. cit. My treatment in *Time and Environment* differs from Perrings's in two major respects. First, it is more intimately informed by thermodynamic considerations in its discussion of the mutual conditioning of contiguous production processes; second, it focuses more overtly on the interplay of interior/exterior components of the inherent indeterminacy of complex system evolution. On open systems and thermodynamics more generally, there is a large literature of very uneven

quality. Some of the more rewarding contributions are: Prigogine and Stengers, *Order out of Chaos* op. cit.; Edgar Morin, *La Méthode: I. La Nature de la nature; II. La Vie de la vie* (Paris: Seuil, 1977, 1980); Herbert Simon, "The Organization of Complex Systems," in H. H. Pattee ed., *Hierarchy Theory: The Challenge of Complex Systems* (New York: George Braziller, 1973); and Martin O'Connor, "Entropy, Structure, and Organizational Change," *Ecological Economics, 3*, 1991, pp. 95–122.

21. This terminology is purpose-invented (see my "Physical Functioning and Economic Control," and my *Time and Environment*, op. cit.). No particular labels seem to exist in the scientific literature, although the distinction between the two cases is pivotal to the whole "self-organization" concept in the far-from-equilibrium thermodynamics (see Prigogine and Stenders's "The New Alliance," *Scientia, 112,* 1977). There is, however, a close parallel with Perrings's distinction between a weak and strong environmental assumption in ecological economics, mentioned below.

22. The term "complexity" is borrowed from Isabelle Stengers, "Complexité," in I. Stengers, ed., *D'une science à l'autre* (Paris: Seuil, 1987). Jean Baudrillard, in *Les Stratégies fatales* (Paris: Grasset, 1983), talks in a related way of the enigmatic genius of the object. What is crucial here is that an axiomatic status is accorded to a property that is widely observable but downgraded as nonfundamental in reductionist and determinist perspectives. This is the autonomy of system components (which may be physicochemical systems, organisms, self-conscious subjects, or societies) as an origin of distinctive behavior and creativity. In the "complex" view of reality, the properties an element displays are not deemed intrinsic and immutable to the observed "object" itself. Rather, the discernible element displays are not deemed intrinsic and immutable to the observed "object" itself. Rather, the discernible components together with their properties "emerge" and are manifested within a collective regime of activity. Objects and properties are the coeffects of the totality of their interactions. A given element can only be understood in terms of its interbeing with the rest of what is (which is, in first approximation, the object's environment). For a systematic critique of ecological economics based on this epistemology, see my *Time and Environment*, op. cit.

23. Perrings, *Economy and Environment*, op. cit., p. 4.

24. Anthony Wilden, *System and Structure*, 2nd ed. (London: Tavistock, 1980), p. 360.

25. Perrings, *Economy and Environment*, op. cit., pp. 4–5.

26. On the hierarchical and mutable character of physical laws, see John Wheeler, "Genesis and Observership," in Robert E. Butts and Jaakko Hintikka, eds., *Foundational Problems in the Special Sciences* (Dordrecht: D. Reidel, 1977) and Bernard d'Espagnat, *Une Incertaine Réalité* (Paris: Gauthier-Villars, 1985). On the death of an individual organism as an ontological event, see Umberto Maturana and Francisco Varela, *Autopoiesis and Cognition: The Realization of the Living* (New York: D. Reidel, 1980). On themes of mutability more generally, I am influenced here by Michel Serres, *Genèse* (Paris: Grasset, 1982); Jean Baudrillard, *L'Échange symbolique et la mort* (Paris:

Gallimard, 1976) and *Les Stratégies fatales*, op. cit.; Castoriadis, *L'Institution imaginaire*, op. cit.; and Latouche, *Le Procès de la science sociale*, op. cit.

27. Underlying these assertions is a more embracing metaphysical position that I cannot spell out fully here. Indeterminacy, and incomplete predictability are regarded positively as inherent characteristics of reality, especially of what is distinctively *social*. To say this gives an immediate and positive status to the diversity of human experience of action and of intention, although at no stage disallowing the search for "determinants" of observed trajectories and events. It permits us to insist on conflict and change not as merely epiphenomenal, but the stuff out of which history is made. Such assertions seem in one respect apodictic. The philosophical justification for raising mutability to the status of an ontological, as well as methodological, precept hinges on arguments of fecundity and internal coherence, on the one hand, and critiques of the logical difficulties of alternatives (notably, the lack of demonstrable grounds for naturalist or other determinist stances) on the other; see, for example, Latouche, *Le Procès de la science sociale*, op. cit., and G. Reuten and M. Williams, *Value-Form and the State: The Tendencies of Accumulation and the Determination of Economic Policy in Capitalist Society* (London: Routledge, 1989). For all that, propositions for or against determinacy seem finally undemonstrable. Adoption of any particular metaphysical position has unavoidable deontological correlates, that is, being coconditioned by the interests and values served, and, indeed, made conceivable by a particular epistemological stance.

28. Anthropological work such as that by Claude Lévi-Strauss and Marshall Sahlins makes this point clear. Charles Perrings (*Economy and Environment*, op. cit., p. 81) follows Godelier in pointing out that the "irrationality" ascribed in much development literature to so-called traditional societies, is only so by reference to capitalist rationality. The benchmark could very justly be inverted. Serge Latouche suggests that in order to comprehend contemporary underdevelopment and its possible denouements, it is necessary to do just this; see his *Faut-il refuser le développement?* (Paris: Presses Universitaires de France, 1986), and *L'Occidentalisation du monde* (Paris: La Découverte, 1989).

29. Jean Baudrillard, *L'Échange symbolique et al mort*, op. cit., p. 246.

30. See Serge Latouche, *La Planète des naufragés* (Paris: La Découverte, 1991), English translation by Martin O'Connor and Rosemary Arnoux: *In the Wake of the Affluent Society* (London: Zed Books, 1993).

31. Marx, *Capital*, p. 171.

32. Baudrillard, "Design and Environment," and *L'Échange symbolique et la mort*, op. cit. The "illegible ambivalence" is an implicit parody by Baudrillard of Laplace's omniscient demon who is capable of reading from any momentary state of the universe, the motions past and future of the greatest bodies and those of the slightest atom.

33. James O'Connor, "Capitalism, Nature, Socialism," op. cit., pp. 25–38.

4

Ecological and Economic Modalities of Time and Space

Elmar Altvater

1. Introduction: Homogenizing Time and Space

Social and ecological processes unfold historically through the dimensions of space and time. In this unfolding, numerous time scales and spatial scales are relevant—from the glacial advances and retreats, to the lunar cycles and the motions of the planets around the sun, to the time taken to prepare and cook a meal. Capitalism tries to regulate historical, social, spatial unfolding according to its abstract "value form"—the supposed commutativity of all values in exchange. The mismatch between the abstract "equal exchange of equivalents" that Marx identified as the key feature of capitalism, and the complex material temporality of social–ecological processes, is a major source of contradiction.

For simplicity, and as a first approximation, we can refer to real historical unfoldings as *ecological* modalities of space and time, and to capital's attempted regulation (codification and control) of these processes as *economic* modalities. The purpose of this chapter is to explore the contradiction between these two modalities. First, I will look at capital's attempt to *homogenize* time and space, indeed (if it were possible) to collapse or abolish space–time altogether. Since this project is manifestly absurd, the real question is what is being occulted

The author wishes to thank Martin O'Connor and James O'Connor for their editorial assistance. An earlier version of this chapter appeared in *CNS*, *1(3)*, no. 3, Fall 1989, pp. 59–70, abridged in translation by Michael Schatzschneider, from *Prokla*, 67, June 1987.

behind capital's reductionist approach to space and time. So I will explore the real and irreducible dimensions of ecological space and time—of thermodynamic irreversibility and of differentiated social–spatial structures—in order to frame the possibility of effective social–ecological reform.

If physical activity could be limited to an infinitesimally small period of time, the concept of physical space would become meaningless. But, since real-world activity "takes time," space coordinates constitute the frame of reference for all social and material activity. Thus an economy without space and time exists only in neoclassical models of "pure economics," and its theoretical relevance remains limited precisely because of this heroic feat of abstraction. Yet this model drives social change. In models of markets and commodities, "exchange" is treated as instantaneous. In reality, of course, it takes time not only to produce goods, but to ship, deliver, unload and unpack, and get them to where they are needed. So the period of a particular activity can never be reduced to "zero." This compression of, and abstraction away from, real space and time is, nonetheless, the aim of capital. To shorten the circulation time of capital is a principle inherent in capitalist development, as a way of increasing the rate of accumulation. With modern transport and communications technologies, the meaning of space defined quantitatively (i.e., distance) and qualitatively (i.e., physical relief and concrete characteristics of social structures) is greatly diminished; space is measured only in terms of the (ever to be decreased) time and expense of crossing it. Today, it is possible to travel between Berlin and New York in under 10 hours; and $500 million can be telexed from Singapore to the Bahamas via London by pressing a button, just as if no physical distance existed between these places. With the separation of money from its material form (metal, paper) and its transformation into energetic, or electronic, money, the spatiality of money circulation tends to vanish. Space is overcome with the speed of light.

So time is abridged, making space meaningless. And, conversely, physical space is conditioned in a way that compresses the time of activity. For the acceleration of material transport, time signifies nothing but an ensemble of impediments. Natural, cultural, and social impediments to the circulation of capital must be removed. This "usurpation of space" has as its purpose the removal of impediments to the acceleration of production and transportation activity. This usurpation is simultaneously the "production of space" and the construction of a "second nature."

Physical space must be adjusted to shorten the time period of every economic activity. The logic of shortening the time of economic activity and the removal of qualitative and quantitative impediments in space is precisely the imperative of capitalist valorization, or (in Weber's categories) the "rationality of Occidental world domination." This means tailoring space and time coordinates of activity to the principles of means–ends optimization.

The creation, in this way, of a form-specific spatial and temporal social system of coordinates through the production of material structures and immaterial norms abstracts from traditional, pristine, and "natural" spatial and temporal coordinates. This is the production of capitalist "sociality," that is, of a particular unified perception of time, space, cause, number, and other basic categories of understanding. As Emile Durkheim put it, "Society cannot relinquish these categories to the arbitrariness of individuals without surrendering itself. To live, society needs not only sufficient moral conformity, a minimum of logical conformity must exist, too."[1]

What is the specific character of capitalism's time to which we conform? In Norbert Elias's words, time has "the character of a social institution, a regulator of social events, a mode of human experience; and chronometers are an integral part of a social order which cannot function without them."[2] Chronometers, instruments for the precise measurement of time, were invented only recently. Previously, activity was a measure of time. Now, indeed, time itself becomes the measure of activity.

Thus space and time are social categories. Moreover, societies are defined by their specific normative commitments.[3] This much is taken as given here. What this chapter sets out to address is not the question of the "sociality" of space and time as such, but rather a social principle that endeavors to reduce time intervals by submitting the quantity and quality of space to the principle of acceleration. Of course, it is not possible to abstract entirely from space and time, since all activity as well as all production and consumption presupposes the transformation of matter and energy from forms with which we are endowed to different forms that we as human beings need. In production and consumption, it is impossible to disregard the quantitative and qualitative properties of space and time, and equally hard to disregard the specific (social, material) use-value aspects of particular products and production processes. Even the reduction of economic processes to money payments by systems theory cannot escape this fact without invalidating its own premises. Even communicative information has a material substrate—paper currency or computer networks, for example—that is the result of deliberate and intelligent material and energetic transformation processes. Its consequence in economic communication systems could be abstracted away only if communication could be arranged without information, which is clearly an absurdity.

2. The Contradiction between Economy and Ecology

Production means production of space and production of nature. The results of production (and consumption) manifest themselves spatially as cultural landscapes, buildings, cities, streets, the ruins of nuclear power plants, canals, sewers and smog, deserts, garbage dumps, and so on. As Marx observed, what makes a particular place a hunting ground is the fact that specific tribes hunt in it. What makes a region a mining region is the fact that metal ore is mined there by mining companies. What makes a region an industrial area is the spatial realization of entrepreneurial decisions or state planning whose goals are the establishment of an industrial region. What makes a particular space a recreation area is the destruction of other areas and the translocation of possibilities for the satisfaction of the human need for vacations in a territory defined by its use as a vacation spot.

In short, it is human beings' own socially organized, normatively imprinted, and politically influenced processes of production and consumption that form their regional environments. This has a double character. The production of space is at the same time its valorization. Its production and consumption has both material and value-oriented facets. Therefore, we can say that their spatial and temporal coordinates have ecological and economic dimensions, respectively. The logics of different functions are therefore at work in the same territorial area. This is already evident in the fact that the transformation of matter and energy during the process of production and consumption is a particular physical process, and hence is bound spatially and temporally in particular ways. But the commodity–money circulation process, with its logic of time compression and its destruction of qualitative and quantitative spatial obstructions, transforms production into a moment of global capitalist reproduction, that is, inserts it within the world market. At this point, it is no longer a question of the transformation of specific resources into an exchangeable use value—for example, from iron ore to raw iron. Rather, the staple commodity, while processed under specific spatial conditions, becomes one element of the total amount of raw iron ore on the world market and "compares itself" to the same commodity extracted and produced under completely different spatial conditions.

The heterogeneity of physical transformation in real space and time—that is, the particularity of materials, place, and ecology—is at odds with the axiom of general comparability in the world market-place imposed by capitalism. Competition in the world market forces capital to converge toward some "average" spatial conditions.[4] The specificities, singularities, and particularities of spatially bound production and consumption are in this way "equalized." Countries and landscapes lose their unmistakable characteristics and transform themselves into segments of world market circulation and global communication. Local, regional, and national particularities of communication (eating, legal forms, traffic regulations, language, and so on) are likely to be regarded as obstructive by capitalism. Patterns of consumption, transported by way of the circulation process and competition in the world market, will express a specific spatial reality reflecting heterogeneous origins subjected to this equalizing pressure. The same is true for patterns of production. Thus the conditions of commodity production and the conditions of capital valorization too must conform to one another for reasons of "competitiveness." The world market manifests itself as pure objectivity, as an impersonal space within which producers and consumers interact freely. In reality, it is a powerful force prompting homogenization, in which natural milieu and social relations alike are transformed according to a "plan" that adheres to the conditions and imperatives of the world market, and that abstracts from specific regions and natural, cultural, and social conditions of reproduction within particular regions.

Competitiveness would not exist if profit were not the aim of production and marketing. But profit is a surplus produced in the production process of specific saleable commodities, relative to those factor inputs that are money-valued. Economic activity, specifically production, is thus determined in two ways. On the one hand, production is nothing but the transformation of matter and energy. On the other hand, it is, in its capitalist form, the creation of surpluses measured in money units, thus abstracted from the complex materiality of the processes and from use value in specific human and ecological contexts.

Transformation of material and energy qualitatively follows certain laws of nature with coordinates defined in terms of physical time and physical space. Georgescu-Roegen[5] has differentiated between two radically distinct notions of time, a differentiation which could be made analogously with the concept of space. First, there is time "T," historical and existential time, describable as a "stream of consciousness" or as a "continuous sequence of moments." Second is time "t," which denotes the time intervals between two activities measured by using a mechanical chronometer. In the passage of time "T" defined as a "stream of consciousness," it is irrelevant when a specific physical process (e.g., the oscillation of a pendulum) takes place. Certainly, one can measure the intervals. However, this does not say anything concerning the particular history or experience. The measurement of time with continuously more perfected chronometers defines time as nothing more than the interval between two activities, regardless of the historical arrangement of the latter. But the reduction of "historical time" (T) to "dynamic time" (t) is not possible without eliminating what is specific to historical experience. Purely mechanical phenomena (or, phenomena described only in mechanical dimensions), says Georgescu-Roegen, do not have a history, properly speaking. In other words, mechanical phenomena are Timeless (*Zeitlos*) though not timeless (*zeitlos*). Correspondingly, only in the mechanistic–temporal "t" sense can processes be unambiguously forecast, and this requires the elimination of all elements of time "T."

One can easily see, in social life as well as in specific scientific and forecasting domains, the growing tendencies toward this dehistorization of time through standardization of event and place—or, more specifically, through focus on supposedly standardized aspects of each event or place. An example is the measurement and timing of a sprint or a ski race. Of course, this uniformization can never be fully realized; yet this is the explicit ideal, and the whole weight and significance of "world records" and world champions depends on this standardization. Regardless of where these activities take place, the only measure of interest is the interval (distance and time) between the start and finish lines. Time "t" is independent of human activities; it is not part of consciousness and is therefore irrelevant in terms of fixing the coordinates of activities.

Thus, under capitalist sociality, a particular logic develops in the space and time coordinates (as in the social and economic coordinates): economic surplus production is guided by the quantitative imperative of growth by way of reducing the time spans of human activities (especially those of production and consumption) and standardizing these activities. It does this by accelerating and transcending the quantitative and qualitative impediments in space in order to compress time, thus turning "T" into "t."

All the same, physical processes in space and time do bear heavily upon human consciousness and activity. In the interval between two activities, that is, during the process itself, entropy has increased and something irreversible has occurred. The temporal aftermath of the activity has left the world changed—and changed in particular ways at particular locations. There are thus two coordinating systems of space and time, which, in the form of two patterns of "functional spaces," are simultaneously at play upon a territorial–social reality.

This is what I meant when speaking of the "contradiction of economy and ecology." The space and time of a society, and the physical time and space of nature, are in no way identical—and this is especially true for capitalism.[6] The logics of their respective functional spaces collides. Ecological crisis can, in many regards, be understood in terms of this collision.

3. Entropy and Scarcity

One facet of historical Time that is of paramount ecological significance is the irreversible character of material and energy transformations. Energy cannot be produced but only transformed from one form to another. The two laws of thermodynamics espoused by Claudius in 1865 are, first, that the universe's energy is constant and material and energetic inputs are always equivalent to outputs; and second, that the entropy of the world strives to maximize itself. Used energy and matter are transformed from the order of unequal distribution to the disorder of equal (homogeneous) distribution, and thus are no longer very useful. As Georgescu-Roegen remarks, "We can use a specific amount of low entropy only once." From the standpoint of the criteria of human use, no transformation of energy or matter is perfectly efficient: a portion is always lost in the form of heat. And if heat is spread evenly, the flow of heat from which energy derives completely stops. Georgescu-Roegen provides us with a convincing example: in comparison with the amount of heat in the ocean, the heat of a ship's boiler is infinitesimally small. Yet the ocean's heat cannot be used, or can be used only with considerable difficulty, while the heat from the boiler can be transformed to propel the ship. As processes of material and energetic transformation, production and consumption too are subject to the law of increasing entropy. This means that the economic system and the tendencies inherent in it cannot be conceptually grasped without reference to their conditionality, that is, to the modes of action of natural laws.

An increase in entropy is inevitable in systems closed to material and energy transfers. In open systems, however, entropy can remain constant or decrease through entropy migration. This reality explains the rampant growth of certain biosystems that are capable of assimilating nutrients and energy from other systems. Even in closed systems the efficiency of energy and matter transformation is variable. This variability is measurable in terms of inputs of energy and matter compared with that part of the output that is useful to human beings; obviously, such a definition of thermodynamic efficiency is anthropocentric. In these terms, then, one can make use of energy and matter more or less efficiently, sparingly, or wastefully, and do this sensibly or senselessly. By the same token, in strict thermodynamic terms, the speed of the inevitable entropy increase can accelerate or decelerate, and also the locations of increase or decrease can be varied.

Studies of plant and (nonhuman) animal ecosystems suggest that, in a general way, the "rate of entropy production" depends on the degree of complexity and diversity of the system. These features determine the scale of nutrient recycling and the necessity for external energy and matter inputs, as well as the susceptibility and responsiveness of the ecosystem to external shocks. Regarding tropical rain forests, for example, it can be shown that the transition to monocultural forms severely increases the injurability of the ecological system as a result of external shocks, and raises the possibility of complete collapse of the ecosystem.

In human social systems, however, the "rate of entropy production" cannot be thought of as a "natural" property (as a function of the complexity, diversity, etc., of the existing structure); rather, it depends very much on what one might call "system intelligence." This intelligence decides how great the rate of exploitation of renewable resources is; if and how processes of substitution for nonrenewable resources are carried out; and how much it is possible to develop the use of time and space to make most effective use of resource supplies and their reproduction cycles. Social conditions and the mechanisms of regulation of society and nature are an immaterial resource—and therefore, in principle, a renewable one—and obviously are decisive for the rate of entropy production. Indeed, ecological economic approaches (e.g., waste recycling and electronically steered savings of chemically bound fossil energy) rest upon social and political options that either increase entropy production or slow down the pace of this irreversible process and limit its spatial dimensions. The question is: Does the potential for system intelligence have restrictions that are embedded in the structure and function of the socioeconomic system itself?

Even if the rate of entropy production can be slowed down, it can never be reduced to zero without complete cessation of organized activity. This truth imposes the reality of opportunity costs for materials and energy use. If sources of energy and raw materials are limited, and these can be used only once (being degraded through use, with the result that full recycling is thermodynamically impossible, or at best very difficult), then appraisal of alternative uses of scarce resources becomes crucial. At this point, economics as the science of the "rational use of scarce resources" enters the scene. Without scarcity, there is no need for economics. If the entropy increase from use of a resource were equal to zero or even negative, the resource could, in principle, be reused time and time again and there would be no scarcity due to depletion. Thus much of economics would lose its purpose. In Georgescu-Roegen's thermodynamics-oriented "bioeconomics," scarcity is axiomatic, on the basis of the second law of thermodynamics. His economics is thus simultaneously an economics of irreversible processes and a science of "natural" scarcity; whereas mainstream economics is premised on the reversibility of economic cycles and approaches the question of scarcity in only an ad hoc manner. Accordingly, says Georgescu-Roegen, mainstream economics is unaware of the thermodynamic basis of its own central category: scarcity. Equally, mainstream economics is heedless of the thermodynamic origins for the necessary production of "waste" with its consequences for pollution, damage to health and habitat, and destructive ecological change.[7]

4. Economics and Time

At this point, the modality of time and the connectedness of past, present, and future material and social processes again confront us. In capitalist economics, the modality of time and the connectedness of these processes have been largely eliminated by the introduction of the concept of "interest," or the discounting of future economic values. In contrast, Georgescu-Roegen's "bioeconomics" stresses that "we must emphasize that every Cadillac let alone every instrument of war means fewer ploughshares for future generations and future human beings, too."[8] The ore of the world's largest ore mine in Carajas in the eastern Amazon region (about 18 billion tons of iron ore with an iron content of 66%) will last about 500 years at an annual rate of extraction of 35 million tons, according to present estimates. Five hundred years is a long, but finite time. Compared to the millions of years that the deposits have existed, it is a tiny time span. For some other raw materials—for example, petroleum—the known and accessible stocks stand to be exhausted much more quickly. Putatively renewable resources such as tropical forests and fisheries are also seriously at risk under current rates and methods of extraction. Earth time, resource time, and human time use different time intervals to measure the time span between past, present, and future. In the connection, the calculations of the Club of Rome pertaining to resource supplies and resource consumption justifiably make sense. They give us a strong impression of the finiteness of resource supplies and their exhaustibility in time, even if their depletion horizon (in terms of present generations) is far in the future. Resources are mobilized by the capitalist economic process in a comparatively short time, and thereafter are available only in quantitatively reduced and qualitatively degraded forms—or, indeed, they are completely and irreversibly "used up," consumed. They then are remembered in the form of their heritage as radioactive wastes, seas of red mud from the production of aluminum, and so on. Indeed, the accumulation of wastes from industrial production and consumption, largely irreversible, may prove seriously disruptive to the global economy long before major sources of raw materials actually run out.

On top of the thermodynamic dimensions of scarcity is superimposed an economic mask, so to speak. Resources can be economically scarce when it is not economically worthwhile to prospect for, develop, and exploit them because of negative cost/gain ratios as measured in money terms. Correspondingly, the accumulation of wastes and degradation of habitats may be neglected, not because the environment as "sink" is nonscarce, but simply because the waste producers are not being required to internalize this degradation as a monetary cost. Scarcity is defined not only by the finiteness of resources and the irreversibility of their consumption, but also economically by the "principle of rationality" that the functional space of economics (i.e., the world market) provides. Paradoxically, scarcity in economic functional space can even lead, in particular conjunctures, to the profusion of resource supply. This occurs, for instance, when "scarce resources" become expensive, and high prices encourage an increase in resource prospecting and exploitation, leading to its abundant availability for a certain time period. Recent examples include the exploitation of North Sea oil deposits, the opening of new oil fields in Texas, and the development of oil surrogates (e.g., produced from sugar cane in the Proalcool program in Brazil). Conversely, a resource field or a program of resource substitution can become uneconomical when resource prices fall. Similar permutations may take place as determinants of waste disposal costs are changed. For example, once chlorofluorocarbons were recognized institutionally as (probable) major causes of atmospheric ozone depletion and a threat in the short or medium term to economic interests, their use was blocked; and new technologies and substances were opened up on the market as substitutes (the major chemical companies did not miss a beat).

Economics as the explanatory discourse for the market system tells us that prices in the functional space of the market signal changing relative costs (opportunity costs of production and use), hence changing scarcities. This is the simple explanation justifying the mobilization of resource supplies (and their suspension) as profitability dictates. Economics, using this scarcity concept and the calculative rationality based on it, thus claims to solve all problems, or to conjure them away.

However, the problems are much more complicated, and this is not only because prices (and interest rates) are subject to the erratic reactions of the world market. More important, the temporal range of economic calculations and the price movements resulting from these calculations diverge sharply from resource times and waste-disposal times. The planning horizon of nuclear power companies, for example, is at most several decades. The half-life period of radioactive waste, however, is some 100,000 years. Economics is, in effect, the science of the *"avant le déluge."* On its banner could be written, *"Après moi le déluge."*

5. Thermodynamics and Economic Surpluses

Understood in terms of the laws of thermodynamics, production is nothing but the transformation of matter and energy, a process in which an available input is transformed into needed output. A "throughput" is thereby produced; but this is a throughput largely external to the grid of gross national product, and whose total effects extend far beyond the temporal and spatial horizons of economic agents. These economic agents (capitalists, entrepreneurs) are not, after all, particularly concerned with transforming matter and energy. In fact, they are largely indifferent to the physical dimensions of such transformations so long as their enterprise remains viable—and specifically, so long as a capital surplus is achieved that permits them to set in motion a further cycle of the transformation process that nets them a surplus.

In this concept of surplus, it is possible to see the abstract circularity of the economic process and the mismatch between this illusion of a self-feeding *quantitative* accumulation and the material realities of this process, namely, the *equality* of inputs and outputs in brute energy and material terms, and the *qualitative* irreversible change in entropy and ecological terms. This contradiction is constitutive of the tortured relationship between economics and ecology in the capitalist mode of production.

Moreover, this contradiction has a specific social dynamic. As Marx showed, the production process is both a labor process in which the transformation of matter and energy is carried out according to the laws of nature, and a valorization process in the course of which an increase in labor value is added to the money capital the capitalist has advanced. This double character of production and reproduction is possible because of the particular social conditions under which it occurs. Its precondition is that labor has been transformed into wage labor and that wage laborers perform surplus labor; that is, they are exploited. In the absence of this specific social form of labor, it would be impossible to occlude the discrepancies between the ecological transformation of matter and energy actually taking place and the refracted representation of this transformation as economic surplus production. Only within the wage form of labor is it possible that matter and energy can be transformed according to an intelligent plan, with a qualitative redistribution of matter and energy flows achieved between social classes in the form of a value flow from labor to capital.

In effect, materials and energy coming under the functional space of economics are subjected to a particular "metric," or mode of measurement, by which they are subsumed under the value form and thus under the money form. They are not valued for any use values they might have in their own setting. Rather, they are considered for their value in capitalist accumulation (production of a surplus), and obtain exchange values in these terms alone. This valorization (monetization) is the precondition for the realization of the "logic" of economic functional space, which, in principle, is twofold.

First, all energy and matter conversions are brought into commensurability with the accumulation project through being measured by flows of money (explicitly with a price, or implicitly as having zero price); henceforth they can be differentiated only quantitatively within this monetary space. This process of money reductionism is obviously very different from a perception of energy and matter transformations in accordance with their qualitative physical (spatial and temporal) and social differentiation—in which terms ecological change will be felt in the world.

Second, this logic of qualitative (ecological) differentiation is repressed, masked behind the economic logic of merely quantitative differentiation. This occulting is what makes historically possible an orientation toward spatial expansion of quantitative accumulation.

These reductive and repressive aspects of the particular form of economic processes are what is neglected in most economic analysis, even when such analysis does try to take into account spatiality and temporality (and hence develop a potential to grasp the contradiction between economics and ecology). Even worse, however, the contradiction is totally ignored when the attempt is made to attribute value to nature, without paying sufficient attention to the particular social form of the "valuing" of nature. Labor only produces value as wage labor;

the pertinence of this formulation (and its attendant theory of exploitation, etc.) is thus premised on particular institutional and ideological arrangements. Which form, then, must nature take to produce value?[9] Ecological economists who try to extend the pricing system to "take nature into account" often fail to take account of the specific form of value in capitalism: such casual extension may simply not work.[10]

The social form (value form) specific to capital thus makes two things possible. First, the quantitative logic of capital valorization abstracts from the qualitative limitation of use-value. In fact, capital finds its fulfillment in technologically transcending, economically externalizing, socially marginalizing, and politically reprimanding all obstacles to the growth of quantitative value (i.e., making profits and accumulation). The possibility of surplus production, and on that basis of the accumulation of capital, gives rise to a social tendency to detach the economic process from all qualitative limitations. The reduction of all qualitative peculiarities to a common denominator that can be expressed in monetary form has made possible the enormous advance of Eurocentric civilization during the last two centuries—but at the same time has destroyed whole social formations and modes of production. Also, during this process the natural environment has been aggressively and powerfully changed, and often degraded or destroyed. Whole mountains have been leveled; oceans have been fished out; species have been exterminated; rain forests destroyed; and huge areas have been transformed into refuse dumps and poisonous seas, lakes, and rivers. All of this has occurred in the name of valorization and growth.

Second, the expansionist pressure inherent in the economic logic of surplus production has a territorial dimension (for production is necessarily always spatial). Surplus production is thus identical to the economic conquest—exploration, development, penetration, and exploitation—of space, in other words, to the "capitalist production of space." At first, space is conquered extensively; subsequently, it is capitalized intensively. What Marx called the "propagandistic tendency of the world market" thus follows from the logic of capital valorization. The effect is the globalization of the "collision" between the economic and the ecological logics—the unpeaceful coexistence everywhere of the contradictory functional spaces of economy and ecology—leaving nothing untouched on the entire globe.

6. Frontiers and Borders

The process of capitalist growth and spatial expansion has no inherent borders, but it is in fact limited by external factors. When "the last tree has been cut, one will realize that one cannot eat money," as the saying goes among West German ecologists. Ecological borders facing capital in its push to widen its frontiers do exist, but they are far away. Up to now, destroyed landscapes in the industrial centers have been transformed into artificial parks by the recreation industry, or bypassed by the offer of trips to the "intact world" of undamaged nature. The precondition, of course, is the monetization of damages, a mechanism which (at least in the rich industrial countries) still functions.

Sooner or later, however, these sanctuaries of "intact nature" will also be destroyed. Then what happens? Are we to recreate in reserves of toxic wastes? To avoid this would require that borders be set up ahead of time—not ecological borders as such, but social ones. As this chapter has shown, it is the social (value) form that not only produces and enshrouds, but also brings to a head, the contradiction between economy and ecology. Immanent borders can be created only if the forms of social reproduction are transformed.

The contradiction between physical and social modalities of the historical time regime and the historical spatiality of capital can be diminished in force (though never nullified because of the irreversibilities entailed by the law of entropy) only by a qualitative increase in social system intelligence, and a removal of obstacles to a conscious and considerate intercourse with nature—obstacles that are today inherent in the social (value) form, namely, the principle of surplus production (profit) and the imperative of expansion (accumulation). We must create social and political border lines before the frontier of capitalist expansion reaches the last ecological border, which would be fatal to the conditions of survival of the human race. Once we realize that a transformation in the *social* forms is what is required, fruitful discussions about ecological reform might begin, which would return us to more familiar issues of social science and environmental reform.

Notes

1. Emile Durkheim, *Die Elementaren Formen des Religiösen Lebens* (Frankfurt: Suhrkamp, 1981), p. 38.

2. Norbert Elias, *Uber die Zeit* (Frankfurt: Suhrkamp, 1984), p. 93.

3. This explains the resurgent interest in the linkages between social relations and spatial structures, which has resulted in a closer connection between social scientists and human geographers at a time when, with few exceptions (Lefebvre, Poulantzas), neither group has theoretically studied territorial space as a social matrix.

4. Obviously, these are "moving averages" as a function of evolution in technologies, shifts in patterns of dominance, obsolescence, fashions, and so on; but the general principle applies.

5. See especially Nicholas Georgescu-Roegen, *The Entropy Law and the Economic Process* (Cambridge, Mass.: Harvard University Press, 1971).

6. Thus Durkheim's understanding of society as a part, indeed, the highest form, of nature is inadequate. He depicts the category of time merely as giving rhythm to social life.

7. On these points, see the chapters in the volume contributed by Juan Martínez Alier, Jean-Paul Deléage, and Martin O'Connor.

8. Nicholas Georgescu-Roegen, *Energy and Economic Myths: Institutional and Analytical Economic Essays* (New York: Pergamon Press, 1976), p. 26. Similarly, Nicholas Georgescu-Roegen, *The Entropy Law*, op. cit., p. 304: " . . . But also every Cadillac produced at any time means fewer lives in the future. . . ."

9. When faced with this question, Georgescu-Roegen simply capitulated. For him, the origin of value devolves to the "enjoyment of life itself"; thus he neglected the specific form of value in capitalism (value was left as an amorphous subjective category). Similarly, in a recent anthology on the subject of "social relations and spatial structures," none of the authors included attempt to tackle the problem of the value form. Space is conceived only as a real substrate of socialization (*Vergesellschaftung*) and not in its relation to nature, where processes are initiated through social action conditioned by thermodynamic laws. After Chernobyl, this is clearly an unjustifiable form of reductionism; the social sciences quite obviously need to enlarge the scope of their analysis.

10. Indeed, it may be that much of the confusion in ecological economics circles over the appropriate "standard of value" or theory of price to adopt, stems from this failure to understand properly the radical gulf between the ecological logic of (irreversible) transformations and the economic logic of commutativity, equivalence-in-exchange, and equilibrium.

5

Social Costs in Modern Capitalism

Frank Beckenbach

1. Introduction

The concept "social cost" is associated with neoclassical economic theory, especially with welfare economics. Why would anyone care about social costs from a Marxist perspective? Are there any costs that are *not* social? My thesis is that social costs are emblematic of the tension between market and nonmarket economic elements—the interrelationship of which, from the standpoint of economic reproduction, is unclear in both neoclassical and Marxist economic perspectives.[1] This chapter attempts to sort out some definitions of social costs and to theorize them in the context of capitalism commodity production. It describes and analyzes the structures that determine social costs and argues that social costs are both monetary and nonmonetary phenomena. I also attempt to arrive at an empirical meaning of this "dubious category" by a systematic classification of social costs. I conclude with some implications for social decision making.

2. Definitions of Social Costs

A dictionary of economics defines *social costs* as follows: "A man [sic] initiating an action does not necessarily bear all the costs (or reap all

The author wishes to thank *CNS* editors Michael Perelman, Bill Livant, and Martin O'Connor. This chapter is based on a paper first presented at the 1986 Round Table on Socialism and the Economy held at Cavtat, Yugoslavia. The version produced here is revised and abridged from that first published in *CNS*, *1(3)*, no. 3, November 1989, pp. 72–92.

the benefits) himself. Those that he does bear are *private* costs; those
he does not are *external* costs. The sum of the two constitutes the
social costs."[2] The assumption of a strict separation of "internal" and
"external" costs across a whole economy is based on the workings
of the market mechanism. The aggregation of both types of costs is
based on the assumption that they both refer to maximizing economic
behavior defined in terms of opportunity costs. However, there is a
problem with such an aggregation as a measure of total welfare,
because it implicitly requires that a dollar represents the same amount
of welfare for each individual; but this asumption contradicts the usual
premise of noncomparability by which the marginal utility of money
is supposed to be different for different economic agents. There are
other difficulties too. The monetization of external costs as a precondition for this aggregation seems to be a self-evident result of the opportunity cost principle. However, the enlargement of this principle to
external costs is necessarily based on the "Coasian view" that externality relations between economic agents can be dissolved into barter
relations—a view which in turn presupposes that there are low transaction costs and that property rights are completely and accurately
defined. In other words, aggregating private and external costs in
monetary terms means that the interaction between decentralized
agents is brought about by an all-embracing, centralized agency that
defines "property."

Contrasting with this view are the investigations (and allegations)
of Kapp, which are guided by the assumption that market relations
create nonmarketable effects and structures. This means that a theory
that includes social costs and environmental disruption must be substituted for conventional theories of market equilibrium. According to
Kapp, "Environmental disruption and social costs have long been
neglected or kept at the periphery of economic theory; they belong
to the more disturbing elements of economic reality which economic
theory since the classics had to set out to analyze with the aid of the
construct of a largely self-regulating equilibrium mechanism capable
of harmonizing micro-economic decisions into a consistent and rational pattern."[3] In conventional theories, the market (commodity) economy is treated as a closed and autonomous sphere of social action,
but this treatment contradicts empirically observable physical interdependencies and causal sequences imposed on the economy by the
environment.[4]

Kapp focuses on the nonmarketable element in the definition of
social costs, that is, those that "are excluded from economic accounting by economic agents . . . and . . . are shifted to other persons or
to society at large."[5] In contrast to our first definition, Kapp seems

to identify external costs with social costs. However, there is a fundamental difference between the two definitions. Kapp is skeptical about the marketability, not to speak of the monetization, of social costs; thus he is also skeptical of the possibility of aggregating private and external costs. In fact, social costs as defined by Graaff do not exist for Kapp, who restricts monetary costs to the domain of private costs. Kapp thus tries to capture the "damage structures" beyond the monetary sphere and the process of shifting lying between individual actions and these structures.

However, even if Kapp's assertion about the restricted scope of market forces and monetization is accepted, his implicit conclusion pertaining to the separability of social costs and monetary costs may be questioned. First, social costs are not confined to destructive effects, lowering of quality, and growing constraints defined in physical terms. They could take the form of compulsory costs insofar as they emerge in the economic accounting of economic agents. Second, even if social costs do not take the form of economic costs, they can, to some extent at least, be included as monetary (income) expenditures. These expenditures can be interpreted either as a compensation for a physical loss or as a requirement for maintaining household reproduction. In any case, these expenditures should be distinguished from other expenditures.

Taking these additional costs and income expenditures together, it can be concluded that in all sectors of the monetary sphere, one part can be separated as social costs. Put another way, social costs have a monetary part that indicates a broader set of processes, events, and states.

In Graaff's definition, social costs are reduced to their monetary (or, at least, monetizable) proportions only. In Kapp's definition, by contrast, monetization is excluded completely. However, there may be both monetary and nonmonetizable aspects to social costs, as, for example, with disputed and uncompensated destruction, damage, or reduction in quality of the physical/physiological environment. Social costs defined this way result from the use of nonmarketable resources in market economies. They can be described in qualitative or in monetary terms (monetization), and in general may comprise a mixture of the two. This monetization can take the form either of an estimation of monetary costs or of an identification of real expenditures required for repair of some physical or physiological damage. Social costs can, however, only partly be reflected in such monetary terms.

It also follows that social costs defined in this way are not truly "external." This is so because, first, social coherence in the economic sense cannot be expressed in monetary terms alone but rather is bounded by the availability of resources from nature and labor; and, second, because monetary terms themselves are not related to an equilibrium/pareto optimum situation resulting from mutual barter exchange based on free will and individual maximization. Thus, contrary to the analytical promises of neoclassical equilibrium price theory, there is no reference point in relation to which any costs can be regarded as "external" (and hence should be "internalized").

3. Social Costs in a Capitalist Economy

Capitalist commodity production is characterized by two different antinomies that are the basis of different types of social costs. The first antinomy is that between private production and a social context resulting from the material interdependence of economic agents.[6] The necessary "groping" for the requisites of social production makes economic allocation a clumsy process that causes manufacturing, sectoral, and local misallocations of labor and cyclical overreactions.[7] In the present discussion, the social costs incurred by such misallocations and overreactions are called "economy-induced social costs." The second antinomy (which is derived from the first) is that between quantities (use values) and prices (exchange values). It includes not only the notion that the interests of economic agents in the preservation and increase in the production of goods are confined to those goods that can be transformed into exchange values, but also the idea that computations of value dominate perceptions and manipulations of the physical/physiological environment. The production of commodities requires noncommodities from nature as sources of raw materials and as absorbers of joint products that pollute, and human labor.[8] In this way, individual production processes are linked informally. In the present discussion, the shifting of costs that is thereby possible is called "ecology-induced" and "labor-induced social costs."[9]

This general description of the problem of social costs needs to be differentiated in two respects. First, economy-induced, ecology-induced, and labor-induced social costs intermingle. The noneconomic linking of production processes results in a microeconomic interest in maximizing the exploitation of human labor power and nature because of the consequences of this exploitation are dealt with in the overall social context. By influencing the cost-return ratio (or the maximization of income), these conditions of exploitation also affect economic allocation. The social costs incurred by the use of nature (ecology-induced) are linked to those incurred by the mode of allocation (economy-induced). Second, the complex construct called "capitalist commodity production" itself must be explained. Total social costs are determined by structures that cut across the distinctions made in traditional theories of capitalism. There are "sociotechnological structures"—overarching production and consumption—that are tailored to a specific social purpose and application. Money, capital, and wage labor are all part of these structures: "money" as an all-encompassing pressure to monetize (particularly in the case of individual reproduction) in the form of a cost–return ratio expressed in monetary terms; "capital" as the interest in dominating wage labor and as a compulsion to grow; and "wage labor" as mass consumption fed primarily by the monetary income of wage earners (and promoted by a variety of credit institutions). Goods and services delivered by the government (through infrastructural facilities) are also part of these structures. In the present account, these complexes are called "reproduction patterns."

There are two reasons why private returns, profits, income, and "utilities" can be maximized through such reproduction patterns via the exploitation of nature. First, some individual costs can be shifted by means of nonmarket distribution systems, natural conditions, the health and well-being of workers (including the additional system of "spouse and family"), and the government. These distribution systems are made up of interlocking subsystems (such as social security and the budget) that have different degrees of dispersion. Second, some of the costs can be shifted within the market system to coproducers and buyers by means of restrictions on allocation and the calculations of costs and prices.

While the advantages of reproduction patterns vary greatly from one social class (and its fraction) to another, there is no question that the advocates of these patterns constitute a coalition spanning all classes. These reproduction patterns are popular because they are advantageous, while their disadvantages are largely shifted to society as a whole, hence generally hidden from the standpoint of the individual. Moreover, the disadvantages of these patterns are found beyond official indicators of economic success (e.g., money income) and can be grasped only through increased awareness of, for example enormous damages, accidents, and catastrophes. Not only is it difficult to take into account advantages *and* disadvantages, but also those groups that are adversely affected often can be identified only after the fact, hence are prevented from articulating their interests. Nor is a change in the behavior of the individual an appropriate remedy for the ills bred by capitalist reproduction patterns—since in this case others would benefit at one's own expense. The burden of "social

progress" is increasingly diffused by the reproduction patterns, hence
the firm and the labor market are progressively dissolved as centers
of social conflict.

4. Reproduction Patterns in the Developed Capitalist Countries

The following reproduction patterns are typical in the developed
capitalist countries.

1. *The growing importance of chemistry for production and consump-*
tion. Restricted availability of natural raw materials has brought about
their synthetic production. Through chemical processes that reduce
the metabolic time required for the production of many commodities,
new basic materials and methods of production are being provided
for nearly all industries. For chemicals as a product of a key modern
industry, the level of interlinkage with the economy as a whole is
above average, a result not only of input/output flows but also of
technology transfers from the chemical industry to suppliers and buyers.[10] The decrease in costs brought about by chemical products and
methods of production is associated with an increase in risks (i.e., the
danger of accidents and catastrophes) for human beings and nature. It
is also associated with social costs stemming from the fact that natural
systems cannot absorb certain kinds and amounts of chemical wastes
and that such wastes destroy natural processes. Differentiation and
growth within the chemical industry is promoted by the economies
of scale involved in chemical mass production and the greater number
of opportunities that such changes create for the use of joint products,
whether or not they are desired. Decreasing microeconomic costs
and increasing social costs are two sides of the same process, and the
chemical industry is one of the main perpetrators of damage to both
the environment and human health. Because dangers result only from
an accumulation of several substances over a long period of time, the
disadvantages for welfare are more hidden than are the instantaneously
available costs advantages associated with chemical products and their
methods of production. At the same time, the pharmaceutical industry
(which is part of the chemical industry) profits from the therapies
and medicines consumed in attempts to cope with the social costs
involved. The chemical industry is supported by government subsidies through health-related welfare aid, research expenditures, and
protection of patents.

2. *The growing important of electronic information and steering systems*
for production and consumption. The characteristic feature of electronic

information and steering systems for production and consumption is an increased supply of information, accelerated information processing, and a standardization of production and work processes. An increasingly large number of such systems are being introduced in private production and consumption, as well as in public communications. Their microeconomic attractiveness is based not only on the fact that they save costs by intensifying work and accelerating the turnover of commodities, but also because they enhance the possibilities for exercising control (hence for reducing corresponding expenses) within the social organizations of firms and management. But this reduction of private (and, to some extent, public costs) also entails social damages: loss of jobs and qualifications; prolonged recovery time; psychic and physiological disorders; and an erosion of the quality of life.

3. *An energy system that treats energy simply as a commodity, the sale of which must be maximized* (even though energy is public managed). The interests of the utility companies are primarily responsible for the waste of energy (e.g., through insufficient use of waste heat and lack of support for the use of insulation), not to speak of the burden on humans and nature through the accumulation of risks (such as those associated with nuclear energy) and heavy pollution (such as that caused by coal-burning power stations). The government supports this type of energy supply through laws, tariffs, and public expenditures (especially in the case of nuclear energy), and through research and development, public goods, and security arrangements.[11] The resulting social costs are shifted to society as a whole through taxes and expenditures for safeguarding health.

4. *A transportation system whose components are characterized by ever greater individualization* (although the system itself is largely determined by the government, e.g., road construction, mandatory liability, and standardization). The private interests of automobile manufacturers and the consumer demand for mobility have led to a destructive process in which the government, as an eager executor of private modes of transportation, destroys both the countryside and the quality of urban life.[12] This mixed system of privately used cars and collectively used roads is both inefficient (because of traffic accidents and under- or overutilization of roads) and inequitable (because the costs to the private users are externalized as a result of products that pollute the air and water, as well as through scrapping, and by the funding through public budgets of the construction and maintenance of roads.) Public transport systems become increasingly unattractive because they come off as losers in the competition for public resources.

5. *A spatial distribution of economic activities that tends toward agglomeration.* Based on private advantage (such as available means of transportation and market shares), economic centers are created, hence a type of income that results solely from increasing scarcity of natural resources (such as land).[13] The ensuing "scarcity prices" bring about a separation of living, working, and leisure time activities, as exemplified by the migration to the suburbs. The diffusion of the corresponding "scarcity income" intensifies pressure for the agglomeration of economic activity. One of the social costs is the increase in traffic (and all its subsequent problems), which is connected with the separation of living, working, and leisure activities. Another social cost is the deterioration of the quality of life in the cities, which is brought about by noise pollution, urban sprawl, and poor living conditions. In addition to the creation of local economic centers, the spatial distribution of economic activities as a whole is marked by uneven regional development, with some regions growing and prospering as other regions decline and stagnate.[14] Costs of agglomeration and costs of pauperization are the two components of the sociospatial costs. Both regional polarization and local agglomeration are promoted by public goods (especially for industry), taxes, and subsidies.

5. The Economic Significance of Reproduction Patterns

The shifting of costs from the individual economic agent to society is common to all of these reproduction patterns. This means that it is possible for capitalist firms to increase their rate of profit (in relation to sales volume) by making use of these reproduction patterns. However, this requires investment spending (e.g., in the use of chemical processes, electronic control media, the energy system, and roads). Hence, the exploitation of human beings and nature is manifested as an increase in the means of production, raw materials, and energy, for a given quantity of commodities, while the corresponding quantity of labor decreases. As Commoner has stated, "In the post-war transformation production technologies with relatively high capital and resource productivity and low labor productivity were replaced by technologies with relatively low capital and resource productivity and high labor productivity."[15] This intensified exploitation of nature (outside the individual production process) and labor (inside the individual production process) is accompanied simultaneously by environmental pollution and unemployment. The microeconomic condition determining the introduction of technologies drawing on these reproduction patterns is that the former will reduce unit production costs (at least temporarily). At least to some optimal point, the greater the scale of production, the likelier it is that such tecnologies will reduce production costs.[16] Hence, economic growth, increase of social costs, and rising profit rates are only different aspects of the same process.

Given the existence of social costs, the aggregation of microeconomic accounting is no longer sufficient for macroeconomic accounting. Moreover, microeconomic outcomes and macroeconomic outcomes can move in opposite directions. As Rohwer, Kuenzel, and Ipsen note:

> In reality . . . the application of new production methods always implies a decision to invest in and use external production conditions to which the market has not assigned a value. Indeed this fact is not ignored in private investment decisions (as it is in the "choice-of-technique models" usually discussed). To minimize private costs, it is essential to use all available "costless" social and natural conditions of the production process. To put it differently, in a capitalist mode of production with given input/output price structures, the cost criterion leads to a selection of the technology that minimizes costs by allowing the extensive use of external conditions.[17]

Generally speaking, a choice of technique designed to maximize the rate of profit, or oriented to microeconomic cost accounting, will produce perverse effects in several ways. First, plants and technologies that decrease microeconomic costs but increase social costs will be built or introduced; second, plants that are profitable macroeconomically but not microeconomically will be shut down; third, certain kinds of plants and technologies will be avoided because, while they decrease social costs, they do not decrease microeconomic costs.[18]

Providing a sort of "investment climate," the reproduction patterns are the bases for technologies and, at the same time, the "tracks" for shifting the resulting costs to society at large.

For most economic agents, social costs constrain decisions. Part of their income must be spent on compensation for informally incurred damages and losses (as private expenses, taxes, premiums, or some other form). Their social function as wage earners or producers of specific goods and services thus may be undermined. Additionally, these expenditures induced by social costs will tie up resources in the "repair industries." If social costs (and their corresponding expenses) are passed on to "society" as a component of prices, a large number of prices will rise, with how high depending on how they are interlocked. This social burden increases with the size of the expenditures induced by social costs.

6. Classification of Social Costs

At this point, it is useful to develop an empirical classification of the most important kinds of social costs. Economic agents do (indeed must) react to social costs. Nonmarket distribution makes it difficult to ascertain damage costs, so costs of evasion, planning and supervision, repair, and prevention also should be taken into account as indicators of actual social costs. Table 5.1 offers a picture of all these factors. Social costs can be identified through an evaluation of the damages listed in column (A). Columns (B), (C), and (D) list reactions of the economic agents to these social costs. Either because it is easier to express these reactions in terms of value, or because they are in fact cited as monetary expenditures, they can be taken as indicators of social costs. Finally, the costs of converting to modes of production (or reproduction patterns) entailing fewer social costs are listed in column (E). Inasmuch that monetization is possible,[19] it stands to reason that this "inequality" says that, for society as a whole, the ceiling on expenditures induced by social costs is given by the cost of overall damages. The ceiling on the introduction of new production methods or reproduction patterns is given by the sum total of induced expenditures incurred.

This table of social costs can be adapted to specific lines of industry or reproduction patterns. Because reproduction patterns span all sectors of the economy (firms, private and public budgets), many cases of intersectoral and intrasectoral shifting of social costs are possible.[20]

7. Conclusions

Social costs can be manifested in the forms of the deterioration of the quality of life, polarization of income distribution, and constraints on the decision making of economic agents. These manifestations of social costs can be interpreted as a differentiation of the conditions for exploitation. On the one hand, labor exploitation is accompanied by the exploitation of society by means of the exploitation of nature. On the other hand, income distribution is no longer a sufficient indicator of exploitation. Another criterion for exploitation is the mode of reproduction associated with a given income (the proportion of expenditures that is earmarked for the compensation of social costs).

The preceding reflections are intended to make clear that neither "growth" alone nor the specific use of given "forces of production" can explain the growing importance of social costs. The divergence between microeconomic and macroeconomic success must instead be explained by specific patterns of reproduction and growth whose attractiveness is based on nonmarket cost distribution. The use value of monetary accounting is being increasingly restricted by this divergence.

In the economic system of the Western industrialized countries, efforts and returns to produce goods and services are expressed as monetary costs, and the possible returns are expressed as monetary income. Given social costs, there is an informal nonexchange mechanism governing the allocation of efforts and returns. Macroeconomic accounting confined to monetary accounting is therefore inadequate for judging economic progress.[21] This is the background for the question of "whether the yields of technical and economic progress are justifying the necessary efforts or whether the so-called wealth is only fictitious."[22]

While their full consideration is essential for a genuine balancing of efforts and returns in society as a whole, social costs "cannot be assigned a value by the price mechanism because they are indicators of nonnmarket interdependencies."[23] Part of the problem can be solved by simulating cost and price accounting (e.g., by investigating damage functions and opportunity costs). Certainly, there are limits to such artificial monetization (such as incomplete information and, more fundamentally, a lack of ways to assign monetary costs to damages and to attribute damages to producers and consumers). Moreover, this monetization is questionable in some social spheres because it presupposes an unwanted "radical" reduction of complexity. Taking account of social costs thus has to remain, in important respects, a nonmonetary evaluation process. As Fritsch remarks, "The problem is to find evaluation criteria that are based neither on particular market solutions nor on dictatorial decisions but rather on a politically feasible evaluation of nonmarket interdependencies that is commensurate with the social conditions of production."[24] The decision-oriented elaboration of these criteria requires one to set minimum standards for the quality of life; set priorities; take economic, political, and social constraints into account; and formulate trade-off relations between the various criteria. How this can be done without falling prey to the temptation to compare the incomparable in the pursuit of choice criteria is a question to be pondered. The point is, all these requirements indicate the shortcomings in the decision-making systems of developed capitalist market economies. The social costs are evidence, according to Fritsch, of

TABLE 5.1. Types of Social Costs

	(A) Damage costs	(B) Evasion costs	(C) Supervision and planning costs	(D) Repair costs	(E) Prevention costs
Ecology-induced social costs					
1. Air pollution	Forest blight hurting the lumber and tourist industries Corrosion Damage to agriculture and food production Deterioration of products Health damages	Resettlement; recreation	Costs for research and development Administration for environmental supervision	Afforestation Catalytic converters Decontamination of power plants Cleaning and accelerated replacement of buildings, corrosion protection, restoration of cultural property and historical monuments Therapy, social transfers	Technologies with minimal pollution
2. Water pollution	Fish mortality with consequences for fisheries Damages to tourist industry and food production Deterioration of products Health damages	Water reservoirs, water transport Resettlement	Research and development Administration for water supervision	Filters Therapy, social transfers	Water recycling Technologies with minimal waste water
3. Soil pollution	Destruction of flora and fauna with consequences for agriculture, tourist industry, real estate business, and food production Deterioration of products Health damages	Resettlement, recreation	Water disposal (research supervision)	Removal of contaminated soil Garbage removal Deposits, incinerators Therapy, social transfers	Technologies with minimal wastes Sorting and recycling of waste
4. Extraction of raw materials	Destruction of countryside and regional structures	Resettlement		Recultivation	Substitution of raw material

Labor-induced social costs					
1. Poor working conditions	Lowering of productivity Health damages		Factory and health inspection	Job protection Therapy, social transfers	Humanization of work
2. Changes in employment/ unemployment and in distribution of working hours in life	Training time Health damages Shortening of life expectancy		Administration for labor exchange	Employment programs Early retirement Unemployment subsidies	Redistribution of work
3. Deskilling	Lowering of value production Expenditures for training			Maintenance of qualification, retraining	Broadening of skills
Economy-induced social costs					
1. Unutilized production capacities	Destruction of real capital, underutilization			Subsidies for owned capital Economic policy	
2. Intersectoral misallocation	Overcapacity, undercapacity		Economic administration	Economic structural policy	
3. Intrasectoral misallocation	Barriers to access, dependence on delivery Overcapacity resulting from concentration		Economic administration	Competition policy	
4. Spatial misallocation	Agglomeration, travel, rent increases, health damages Costs of regional pauperization		Administration (including regional administration)	Spatial policy Rent subsidies Therapy, social transfers	
5. Planned obsolescence	Premature substitution	Substitution of products	Consumer protection Inspection of food		Additional costs for high-quality production

how ever-greater nonmarket interdependencies are forcing modern industrialized societies to undertake more and more evaluations and decisions without their having either the appropriate institutions to do that or the objective basis on which to derive adequate criteria for evaluation. Neither the market nor parliament can have this function. Hence, even in such important fields as nuclear power, the development of transportation systems, and growth policy, far-reaching decisions are based on momentary incidental constellations of interests instead of carefully prepared and economically sound evaluations and priorities.[25]

Notes

1. A sketch of the difficulties in classical, Marxist, and neoclassical accounts of social costs, as well as in the best-known account by William Kapp, is F. Beckenbach, "Zur Theorie der gesellschaftlicher Folgekosten," in F. Beckenbach and M. Schreyer, eds., *Gesellschaftliche Folgekosten* (Frankfurt: Campus Verlag, 1988), pp. 13–49.

2. J. de V. Graaff, "Social Cost," in *New Palgrave Dictionary of Economics* (London: Macmillan, 1987), vol. 4, pp. 393–395.

3. K. W. Kapp, "Environmental Disruption and Social Costs: A Challenge to Economics," *Kyklos*, 23, 1970.

4. Ibid., p. 840n. Kapp's critique does not answer the question: How can noneconomic interdependencies be integrated into a disequilibrium explanation of macroeconomic coherence based on microeconomic decisions?

5. K. W. Kapp, "Sozialkosten," *Handwoerterbuch der Sozialwissenschaften*, (Gottingen: Gustav Fischer, 1956), vol. 9, p. 525.

6. Marx called this material interdependence the "division of social labor." This is a shortcoming, as there are interdependencies that are not mediated by labor, for example, the natural conditions of production.

7. O. Bauer, *Kapitalismus und Sozialismus nach dem Weltkrieg* (Vienna: Wiener Volksbuchhandlung, 1931), pp. 184–202; B. Link, *Social Costs* (Bern: Herbert Lang, 1969), pp. 36, 37–39. It is not the neoclassical "frictions of adaptation" that are the issue here but rather the formation of the economic structure in general.

8. The quantity of these polluting joint products is a function of the technical and economic conditions of recycling (K. Marx, *Das Kapital*, vol. 3 [Hamburg: 1894]; reprint, East Berlin: 1966, pp. 111–112). It may be that the increasing production of these substances indicates an asymmetry between the capacity of natural catabolism and the capacity to synthesize joint products.

9. By contrast, the orthodox Marxist view holds that social costs are the result of the commodity-like character of the elements of production. See, for example, Bauer, *Kapitalismus*, op. cit., p. 175, on labor power. In the present chapter, it is asserted that social costs are related to the limits of

the commodity character in relation to the elements of production, and thus also to that noncommodity element, labor power.

10. W. R. Streck, *Chemische Industrie* (Berlin: Duncker and Humblot, 1984), pp. 26–27.

11. P. Hennicke, "Energiepolitik im Umbruch," in Projektgruppe Gruner Morgentau, ed., *Perspektiven oekologischer Wirtschaftspolitik* (Frankfurt: Campus Verlag, 1986).

12. W. Sachs, *Die Liebe zum Automobil* (Reinbek: Rowohlt, 1984).

13. This is an illuminating example of the consequences for nature being expressed in prices, according to neoclassical economics.

14. For the Federal Republic of Germany, see M. Krummacher et al., *Regionalentwicklung zwischen Technologie boom und Resteverwertung* (Bochum: Germinal, 1985); for the United States, see L. Sawers and W. K. Tabb, *Sunbelt–Snowbelt: Urban Development and Regional Reconstruction* (New York: Oxford University Press, 1984).

15. B. Commoner, "The Environment and the Economy," unpublished manuscript, 1985, p. 7. It remains to be investigated if there is a trade-off between the exploitation of nature and the exploitation of labor and if epochs of capitalist development are dominated by one mode of exploitation or another.

16. Ibid., pp. 11–12.

17. G. Rohwer, R. Kuenzel, and D. Ipsen, "Marx und die gegenwaertige Akkumulationskrise—Ueberlegungen zur Theorie der Profitratenentwicklung," *Prokla*, 57, 1984, p. 30.

18. Bauer, *Kapitalismus*, op. cit., p. 172.

19. Although complete monetization is not confined to the scope of market mechanism, it is impossible in a practical sense, and profoundly ambiguous (and indeterminate) even at a purely theoretical level.

20. See B. Fritsch, "Zur Theorie und Systematik der volkswirtschaftlichen Kosten," *Kyklos*, 15, 1962, p. 240. For a complete classification of intersectoral externality relations that takes into account the differentiation between external advantages and disadvantages and between causes and effects.

21. Fristch, "Zur Theorie," op. cit., p. 197.

22. Link, *Social Costs*, op. cit., p. 25; Bauer, *Kapitalismus*, op. cit., p. 182.

23. Fristsch, "Zur Theorie," op. cit., p. 193.

24. Ibid.

25. Ibid., p. 198.

6

Nature, Woman, Labor, Capital: living the deepest contradiction

Ariel Salleh

1. Women's Resistance: An Embodied Materialism

In the 1960s a social movement began to emerge around actions as diverse as women's legal challenges to giant nuclear corporations in the United States and tree-hugging protests against loggers in northern India. These actions signaled a new politics, grounded materially in understandings that come from women's everyday work to meet life needs. Despite cultural differences between the various actors, such actions reflected a common intuition that somehow the struggle for a "feminine voice" to be heard is connected with struggle for a nurturant attitude toward the living environment. The fractured term "eco-feminism"—occurring spontaneously across several continents during the 1970s—encapsulated this double-edged political concern. By the late 1980s ecofeminism was expressing an explicit challenge to the transnational structure of capitalist oppression, that is, to a global economy in which so-called advanced societies are rapaciously dependent on the resources and labor of an "undeveloped other."

This chapter draws together ecofeminist diagnoses of capitalism, looking at the way in which women's labor experiences house both "grounds" for an ecopolitical critique and actual "models" of sustain-

Ariel Salleh is an ecofeminist activist, whose book on convergencies and contradictions between socialism, feminism, and ecology will be published in 1995 by Zed, London. She is a convener of the Women's Environmental Education Centre in Sydney, Australia, and occasional visiting scholar in the Environmental Conservation Education Program at New York University.

able practice. Section 2 discusses the low *value* accorded to "nature" and to "women" under capitalism, and how the exploitation of each intensifies with economic globalization. The *nature–woman–labor nexus* is examined more closely in sections 3 and 4, where it is proposed that this should be considered a *primary contradiction* of capitalism. Section 5 introduces a depth analysis to show how living within this contradiction activates resistance—historical *agency*.

My argument highlights distinctive structural issues that women face under a capitalist mode of production. While economic "growth" appears to have brought material benefits to some men and some women in the North, in another sense it can be said that almost all women inhabit the South. The annexation of women's work is reinforced with industrialization and consumerism, whether by computers, labor-saving gadgets, or new reproductive technologies. Meanwhile, in "developing" regions expropriation of farmlands for commodity markets, technocratic "green revolutions," and now corporate gene patenting undercut the very means of women's labor for subsistence.

Continued capital accumulation and the expanding hegemony of transnational operations deepens nature's and women's subjection. This is not to say that capitalism has been the only source of such oppression, nor to argue that capital does not also exploit men. Rather, it is to make visible something largely unspoken in existing theoretical analyses by pointing to what is unique about women's environmental responses. For the fact is that in at least four ways women's relation to "nature," and therefore to "capital" and "labor," is constructed differently from men's.[1]

The first such difference involves experiences mediated by female body organs in the hard but sensuous interplay of birthing and suckling labors. The second difference follows from women's historically assigned caring and maintenance chores that serve to "bridge" men and nature. A third difference involves women's manual work in making goods as farmers, weavers, herbalists, potters, and so on. The fourth difference involves creating symbolic representations of "feminine" relations to "nature"—in poetry, in painting, in philosophy, in everyday talk, and so on. Through this constellation of labors, women are organically and discursively implicated in life-affirming activities, and they develop gender-specific knowledges grounded in this material base. As a result, women across cultures have begun to express insights that are quite removed from most men's approaches to global crisis—whether these be corporate greenwash, ecological ethics, or socialism.

Far from being premised on simple polarities of masculine and feminine, culture and nature—as some critics of ecofeminism have implied—this standpoint actually rests on a dialectical deconstruction of these received dualisms. It is a political commitment grounded in women's economic marginalization and the painful awareness of contradiction or nonidentity that their place in the *nature–woman–labor nexus* gives them. The strategic privileging of the marginal voice here is thus justified empirically, rather than by some trans-historical or "essentialist" claim. Formulated as an embodied material-ism, ecofeminist politics gets at the lowest common denominator of oppressions. As such, it opens up new possibilities for dialogue be-tween classes and social movements resistant to capital.[2]

2. Value

> My mother used to say that the black woman is the white man's mule and the white woman is his dog.[3]

For ecofeminists, capitalism appears as a modern form of patriarchal relations, in which most women experience a social reality very differ-ent from their brothers in capital or labor. Relatively few women possess assets in their own right, and the majority of women are "not quite labor" either. Even U.N. figures cannot hide the global scandal of feminine marginalization, for women own less than 1% of all property and do two-thirds of the world's work for 5% of all wages paid.[4] In fact, women's place in this predatory system is notionally somewhere between a "natural resource" and a "condition of produc-tion."[5] Either way, women are treated as an economic "externality," just as they have been a historical externality in bourgeois liberal political institutions.[6]

A glance at the post-World War II conjuncture gives substance to these claims. In her classic statement *The Global Kitchen*, activist Selma James points out that

> in the United States in 1979, only 51% of adult women were "in the [paid] labor force," 48% in China and France; in Latin America only 14% of the total female population was counted as workers in 1975. In Britain, 40% of women are in the paid labor force now.[7]

New Zealander Marilyn Waring updates the indicators in *Count-ing for Nothing*. But while a burgeoning service sector in the North, and an explosion of free trade zones in the South, shifts the statistics around a little, the basic character of this female exploitation remains unchanged by globalization and the workplace restructuring that comes with it. Women swell the ranks of part-time, contract, and seasonal positions, without security, advancement opportunities, or retirement benefits. Maternity leave and work-based childcare programs are a rarity. This entrenched gender division of labor is so fundamental to the fabric of capitalist society that, 20 years after a "sexual revolution" and installation of affirmative employment schemes, even salaried women in the industrialized nations typically receive only two-thirds of an average man's wage. More significantly, the greater portion of women's labor is left out of gross national product (GNP) calculations altogether.

Yet a housewife in the "developed" world often puts in at least 70 unsalaried hours a week—almost twice the standard Australian working week of 40 hours. Using subsistence skills, she produces "use value" by cooking, sewing clothes, cleaning, house maintenance, gardening, and so on. Nonmetropolitan women in the South grow the bulk of their community's food. Then there are the intangible obligations of women's open-ended labor role: tending children, comforting the aged and sick, providing ego repairs and sexual relief for the man in their lives, and possibly the labor of childbearing consequent to that. Mary Mellor from the United Kingdom describes all this as putting in "biological time."[9] In addition, many middle-class women take on a heavy round of voluntary commitments, for example, PTA, Amnesty International work, or resident action campaigning. Migrant and refugee women use extra energy absorbing new strains on the family and rebuilding community, often after a full day in the fields or on the assembly line.

The unpaid services—"labors of love"— that women give out under capitalism can, in principle, be remunerated: examples are prostitution, fast lunch counters, professional laundry. This shows that there is no natural necessity to organizing the economic system in this way, only capitalist patriarchal convenience. As Selma James notes,

> The woman who cleans a house is not "working," but the military
> man who bombs it, is. Further, . . . the work of the same woman,
> if hired by her husband . . . would pop into GNP.[10]

The paternalism of capitalist economic arrangements is such that even when women's domestic labors are recompensed in the form of supporting mothers' pensions or benefits for elderly care, these payments are perceived as "a gift" of the state, charity, or welfare, and never as an "economic exchange" transacted between free citizens, as in the contract between "labor" as such and capital.

Using standard economic criteria, one can easily demonstrate the significance of women's contribution to the capitalist economy. James,

Waring, and others including Hilkka Pietila, an ecofeminist from Finland, all substantiate that if we were to allocate domestic hours to standard job categories, apply the going wage, and then total everything up, we would find housework constitutes around one-third to one-half of GNP. The Australian Bureau of Statistics estimates household labor as equivalent to 52–62% of GNP.[11] But if domestic labor were to come in from the cold like this, giving "women's work" a place in "the formal economy"—with the massive redistributions in incomes and patterns of economic opportunity this would entail—would it mean that women themselves were more highly valued by society? Most feminists doubt it, for women's oppression is not simply economic. In any case, to advocate such reform is to presume that the capitalist system at large and the patriarchal family as a microcosm within it are institutions worth preserving.

In sum, women's work makes accumulation possible for all kinds of men, and the "surplus" women generate is quite crucial to the operation of capitalist patriarchy. This statement is relatively uncontroversial, at least among women. During the 1970s and 1980s an extended exchange was carried on among socialist feminists concerning the interaction of capitalist and patriarchal systems.[12] Agreement over the precise scholastic formulation covering women's subordination was not reached. The overlap of female exploitation with ethnicity, race, and the North–South axis was barely touched. However, by broaching the "the nature question," ecofeminists are now reframing the entire debate.

By introducing the nature–woman–labor nexus as a fundamental contradiction, ecofeminism affirms the primacy of an exploitative, gender-based division of labor, and simultaneously shifts the analysis of all oppressions toward an ecological problematic. While liberal feminists may be content with receiving nothing more than equality alongside men in the existing system, ecofeminists are concerned about global sustainability as much as gender justice: in fact, they see the two as intrinsically interlinked. For example, Berit As from Norway argues that economic growth in a male-oriented economy only adds new burdens to women's lives.[13] Money that might sustain women breadwinners goes instead into armaments, six-digit executive salaries, and a paper whirlwind of speculation. Under capitalist patriarchy, it is men in government, business, unions, academia, and international agencies who hold most decision-making positions, and who set priorities that are comfortable for them. The presence of a few female executives in the corporate hierarchy will have little impact as long as masculinist priorities remain unchallenged. One may consider the uncritical contribution of women economic advisers to national governments or the Organization for Economic Cooperation and Development in promoting the General Agreement on Tariffs and Trade, heedless of the intensified exploitation that deregulated markets bring.

Yet it is not just women's livelihood at stake here; the natural environment is equally externalized and decimated by these priorities. The structural intertwining of women's exploitation with the depredation of nature is illustrated at "development's" every turn. Ethiopia suffers desertification and famine as land is taken out of women's hands by men who would "render it profitable." In the United States, women working for electronics corporations are exposed to toxic contaminants of skin, lungs, and nervous system, and they suffer fetal damage. Import of tractors to Sri Lanka degrades soil and water, and forces women to pick cotton twice as fast, in order to keep their wages at the same level. Following engineering failure at Chernobyl in the Ukraine, mothers across Europe pick up the community health costs of nuclear radiation. Sex tourism, a male-organized and male-oriented skin-trade, balances "foreign exchange" in the South, as debt accumulates from the rush for ecologically disastrous masculinist status symbols like weapons, hydroelectric dams, and oil. Living things are expendable for capitalist patriarchy, which does not value what it does not itself produce.[14]

Sisters North and South have more in common than many think; and that commonality increases as the so-called "level playing field" of the economists expands across North and South. The gender rule applies cross-culturally, and for women it reads "Maximum responsibilities, minimum rights." Hence, while technology transfer from core industrial powers—in particular the United States, Germany, and Japan—introduces an era of neocolonialism to the periphery, "development" also heightens the subsumption of women's work. Vandana Shiva portrays this trend as the result of an implicit pact between advisers from the North and local elite men, the upshot being "modernization" projects and structural adjustment programs passing the costs of "economic" growth down the line to women, and then to nature.[15] Village girls become silicon slaves, while the erosion of traditional land use rights with cash-cropping strips their mothers of cultural autonomy and economic control over their means of production.

In India, a culturally sustainable woman–nature metabolism has been undermined by imported scientific techniques that impose an inappropriate linear reductionist "logic" on the cyclic flows of nature. Shiva writes:

The forest is separated from the river, the field is separated from the forest, the animals are separated from the crops. Each is then separately developed and the delicate balance which ensures sustainability . . . is destroyed. The visibility of dramatic breaks and ruptures is posited as "progress."[16]

Indigenous women's expertise developed over thousands of years—knowledge of seed stocks, the water-conserving properties of root systems, transfer of fertility from herds to forest, home grown medicines and methods of contraception—are lost. Nature is broken; human needs go unfulfilled; societies and cultures disintegrate as rural men leave families for the city lights and promise of a wage. Meanwhile, men of the *comprador* class and their World Bank role models publish annual trajectories of "manpower" requirements: engineers, accountants, chemists, whose very skills exacerbate the entropy.

Ecofeminists have long argued that an identification of women with nature defines women's work in the North as well as the South. Take the complex of tasks that housewives perform under capitalist patriarchy: providing sexual satisfaction, birthing and suckling children, carrying the young about, protecting their bodies and socializing them, growing and cooking food, maintaining shelter, sweeping floors, washing and mending clothes, dealing with garbage—and these days recycling it. The common denominator of these activities is a labor "mediation of nature" on behalf of men, which function continues despite legal recognition of "female equality" by nation-states. Such formalities are incidental to the underlying "accord" between governments, capital, and labor, guaranteeing each man his own piece of "the second sex."

3. Contradiction

Women's traditional positioning between men and nature is a primary contradiction of capitalism, and may well be the deepest, most fundamental contradiction of all. In anthropological terms—shaped by androcentric interests—women's bodies are treated first as if they were a "natural resource," with the uterus as organ of birthing labor being the material origin of "formal labor" as such. The time-honored European imagery of Mother Nature and the ancient Indian notion of Prakriti are certainly more than metaphor. But under the scientific hegemony of capitalism, their celebration of women's potency is greatly diminished in favor of a celebration of men's productivity aided by technology.

In European mythology, discourses on produced wealth, nature, and labor take their distinctively modern shapes from around the 17th century, as medieval religious thought is transposed into a secular view of nature. Land is seen as the mother of wealth, and labor as its father.[17] The entire world is a vast pool of resources, available to men in common as a matter of Divine Providence. But wealth properly speaking is a product of men's labor. Every man, says John Locke, "has a *property* in his own *person*," and so "the *labor* of his body and the *work* of his hands, we may say, are properly his." If, in the providential sense, Nature is "the common mother of all," conversely it is through labor that an individual appropriates the fruits of Nature to himself, "so they became his private right."[18] As far as labor is concerned, it is a man's world. One infers that women's domestic and reproductive labors are furnished as "gifts" to men, in return for personal protection in the private sphere.

While women's bodies under capitalism have never come to obtain a rent as land does, they are nonetheless "resourced" for free by capital to provide ever new generations of exploitable labor. Consequently, given that women are really human beings, a profound antagonism is set up between "woman" as objectified reproductive matrix and women as subjects of history in their own right. Currently, this tension is expressed in the form of a reproductive rights debate over abortion, for example, and in the form of arguments concerning the issue of paid surrogacy and the possibility of an "industrial contract" for childbearing in a "value-added" world. How the line may be drawn between woman as "natural resource" and woman as "not quite labor" appears to be infinitely flexible.

In addition to being a "natural resource," women using hands and brain in caring labor become subsumed under capitalist patriarchy as "conditions of existence," in the sense of *oikos* or habitat, necessry for creative human productivity to take place. Women's bodies are utilized by working men to provide a taken-for-granted daily infrastructure, enabling performance of the male work role. The fact that men are bothered rather more by the loss of a wife than by the level of their wage demonstrates a wife's value as a "condition of production"—sexual, psychological, and economic. At the same time, since women are "not quite labor," they find themselves existing in contradiction with "labor as such," and this is so even when they become paid workers themselves. The tensions between women and "formal labor" erupt within the family and at the workplace, with formal labor backed up by a masculinist trade union movement.

Women are doubly objectified by these two forms of structural violence. Like nature, they are readily available and disposable; and like nature under capitalist partriarchy, they have no subjectivity to speak of. Meantime, as Naomi Scheman observes, men are free to imagine themselves as self-defining—but only because women hold the intimate social world together.[19] Women, really "objects," in a so-called division of labor, have customarily been exchanged between men, father to husband, pimp to client, from one entrepreneur to another. This exchange of female resources may well have constituted the earliest form of "commodity" trade.

Likewise, the children women produce are appropriated and named by men. Moreover, even as women begin to take back control of their fertility from the patriarchal family, so men use new reproductive technologies to wrest control of that "resource" back from them. The latest move on this front is corporate patenting of DNA, whereby the basic building blocks of life itself are formulated as "property rights." And this will cover not just "genetic" interventions in human reproduction such as purported remedies for inherited ailments, but transgenic combinations between animal and plant life as well.

Women also "make goods," for use in domestic shadow labor, and for exchange in peasant agriculture, or as commodities in piecework or factories. Yet these commodities too are usually taken away by men—husbands, middlemen, or transnational management. In her *Patriarchy and Accumulation*, German ecofeminist Maria Mies documents this process of dispossession, and observes that violence pervades every facet of male–female interaction under capitalism. By this means, men are simultaneously agents for capital and for themselves as workers, keeping women intimidated and pliable.[20]

Although the oppression of men by men along class and racial lines is well documented, the extraction from nature and from women's complex of productive capacities long predates the theft of value from a working class. Moreover, nature and gender exploitation subsists through and beneath capitalist abuse of wageworkers, and this process is being deepened with global expansion, despite the modern rhetoric of female emancipation. Socialism until now has tended to place too much emphasis on a theory of the proletariat and backgrounded different forms of social exploitation.[21]

An ecofeminist analysis asserts that the enclosure and privatization of women—the subsumption of women's time, energies, and powers—through patriarchal family and public employment alike, parallels the class exploitation of labor by capital, and at the same time it makes the latter possible. Women's position as "mediator of nature" constitutes a prior condition for the transaction that takes place between capitalist and laboring men—big men and small. In the androcentric discourse of economics, the material contribution of women remains largely unspoken in much the same way that the material contribution of nature is attributed zero value. Women's labor is "freely given," or hidden behind the curtains of domestic decorum. What women do "gratis," whether birthing labor or sustaining labor, is called "reproduction" as opposed to production. Yet, the word *reproduce* here connotes a secondary or diminutive activity, as distinct from the primary "historical act" of production itself. And since reproduction is not recognized as "primary," it cannot be seen to generate "value."

By a symbolic sleight of hand sometimes called "reason," women's work is cheated of a place in a system of accumulation resting on the "surplus" they create. The following poem by Paul Eluard conveys something of the innocence and plausibility of patriarchal naturalism and its deletion of women. In the late 20th century, "Sound Justice" can only be read with irony.

> It is the warm law of men
> From grapes they make wine
> From coal they make fire
> From kisses they make men
>
> . . . .
>
> A law old and new
> Self perfecting always
> From the depth of a child's heart
> To supreme reason.[22]

4. Deconstructing Woman/Nature

To understand how such "reason" works, an ecofeminist analysis of the ontology of capitalist patriarchy is useful. The latter hangs on a classical "logic" of dualisms that penetrates philosophy just as much as everyday talk. The symbolism of these time-honored pairs reiterates the morphology of sex, erases women's humanity, and functions to keep men superordinate to women and to "nature." The domain assumptions of capital's discursive armor are as follows:

- An artificial distinction between "history" and "nature."
- An assumption that men are active historical "subjects" and women passive "objects."
- An assumption that historical action is necessarily "progressive" and activities grounded in nature necessarily "regressive."
- An association of masculinity with the historical order through "production" and association of femininity with the order of nature through "re-production."
- "Valorization" of productive activity and "devalorization" of reprodutivity.[23]

Obviously, it makes no sense to speak as if nature is somehow prior to history, for time is a condition of all existents. But what is also missing in these discursive formations is any reflexivity in understanding the grounds for these constructed categories. In epistemological terms, capitalist–patriarchal thinking simply floats on thin air. The "natural order" can be known only through history, that is, by subjects living within a medium of socially generated languages and practices. Capitalism manages to obscure this historical dimension by the sheer force of its ideological machine—such that people actually do come to believe reality is striated in this way, and universally so. Religion, ethics, economics, and even sociobiology hang on these essentialist dualisms. Some leftist critical thought, and even varieties of feminism, are infected by them too, taking the content of these paired assumptions as given.

While a careful deconstruction of conventional essentialist thought categories is needed, what is undeniably given is the fact that women and men do have existentially different relationships to "nature" because they have different kinds of body organs. But to say this is not to say that women are any "closer" to nature than men in some ontological sense. Rather, it is to recall Marx's teaching that human consciousness develops in a dialectical way through sensuous bodily interaction with the material environment. Just as someone who has no organ of sight may develop a unique awareness, so men and women, differently abled, come to think and feel differently about being in the world as a result of how they can act on it, and how they experience it acting on them, in turn. Here we are talking about a kind of knowledge that is shaped by body potentials.

However, people never know this potential in any pure sense, since bodily activities including labor are mediated by language and the ideological constructions embedded in it. Accordingly, women's sensuous interchange with habitat gets to be shaped in a second-order sense, by assigned roles that force them to "mediate nature for men." Historically trapped in the logic of masculinist reason, women's sensual enjoyment and creative reciprocity with their environment is denigrated as regressive by an artificial and compulsory association with nature. In such labors, women give up the substance of their bodies, experiencing entropy like that which nature suffers in the process of accumulation. Curiously, while the value of their work does not register in national accounts, their deterioration does. So the capitalist state provides a plethora of clean-up programs—for example, battered women's refuges, addiction counseling—that parallel environmental efforts at resource recycling and restoration of toxic lands.

In the discursive construction of gendered labor, mining or engineering by men is also a hands-on transaction with the environment. But such work is typified by the positive side of the symbolic grid, endorsing masculine identity as separate from nature, productive and progressive. By contrast, the language that typifies women's work—"re-production"—degrades her along with nature itself. This pseudo-ontology is legitimated by all the institutions of capitalist patriarchy: church and state, market and trade union, technology and science. Consequently, when women challenge this status quo for a share of male privilege as "labor," they meet ideological weapons like harassment and rough handling in order to "reinstate" their properly feminine status as part of nature. This dynamic is inevitable, since "formal labor" can purchase progress under capitalism only by trading off further exploitation onto women and thence nature, down the line.

Of course, male workers are also abused as "conditions of production" under capitalism, but this is not sufficient reason to neglect the distinctive constellation of women's exploitation. What ecofeminism demands is a fully amplified critique of capital's degradation of "conditions of production," based on a recognition of the *nature–woman–labor nexus* as a fundamental contradiction. The treatment of women becomes abusive when, in the analysis of capitalism, the complex of distinctively feminine labors is seen as somehow auxiliary and sidelined in favor of a historically privileged proletariat. As Giovanna Ricoveri expresses it, only by being open to "difference" can one hope for

> An alliance—or set of alliances—that would not involve merely merging the various political components, nor the standardization of cultures, nor limits on the freedom for every group or tendency to experiment freely, but would rather be a Hegelian "sublation," the creation of a new politics that would contain strong elements of the green, the red, feminism, and so on, but would look like none of these well-established tendencies.[24]

Until the problem of gender blindness in politics is overcome, however, women need to be on constant guard against premature theoretical closure in any new totalization. This is part of why it makes sense to strategically prioritize ecofeminist voices at this point in time.

5. Nonidentity and Dialectics

As an emerging consensus among women, ecofeminism is overdetermined in the structuralist sense. But in order to understand what drives the individual agent behind this new movement, the particular sensibilities brought into play, a deeper materialist analysis is useful—one that overcomes the patriarchal divide between nature and history, an embodied materialism that affirms labor as sensuous practice and also moves beyond this to consideration of an interior dialectic between bodily energies and discourse.

This implies a recognition that somatic states make and unmake subjectivity, and indeed, knowledge. After all, what is the individual subject but a body that carries intention? Abrased by contradictory meanings, this subjectivity becomes an active field. Julia Kristeva, in what she calls semanalysis, posits a special state of apprehension where, under stress, body drives and their ideations disintegrate and reassemble. This matrix of apprehensions—or "*chora*"—is the very kernel of historical consciousness, and is renewed again and again through a multiplicity of cathexes that feed the link between signifier and signified.[25]

Under capitalist patriarchy, women find themselves lodged inside/outside relations of production in a way that is contradictory and unlivable. Daily they are broken on the contradiction that has them "closer to nature." Women are human, but they are still treated by the social system as simple reproductive sites, or as commodities, made use of and exchanged like any other "natural resource." Being "not quite labor," they achieve neither financial nor ideological equality in the work force. Having "no subjectivity to speak of," their voices remain unheard, unless to chorus the masculinist discourse with its dogmatic dualisms, thereby affirming their own diminutive role.

How does a women ever find her way out of this double-bind, let alone come to act for social change? I have argued elsewhere that it is through crisis and moments of nonidentity that she glimpses new meanings in her situation, a hidden political potential behind what is given. This "negative dialectic" rests on a distinction between essence and appearance, where the positives of perception—immediate facts—are merely temporary manifestations, even distortions of an immanent reality or essence yet to be explored.[26]

Sexual abuse and domestic battery, economic and cultural marginalization—these things are enough to fracture a woman's identity. Invalidated by contradictory significations in a world that preaches love but practices exploitation, the feminine object/subject decathects somatic energies that tie her to existing social relations. Becoming free from her historically ascribed "otherness," as a subject-in-process she may begin to predicate an alternative relation to the totality. To paraphrase Kristeva: when the fragile equilibrium of consciousness is destroyed by the violent heterogeneity of contradiction, the body returns to a state of difference, heavy, wandering, dissociated. However, moments of annihilation and decomposition of the sense of subjective unity, moments of raw anguish and disarray, can yield up a new productive unity, so reaffirming the subject as active signification-in-process. It is this kind of personal transmutation that usually grounds an ecofeminist epistemology, though women vary in awareness of such inner processes.

Always in the front line of environmental impacts, eroded as nature is, a woman's dis/location may eventually shatter the taken for granted perception of capitalist patriarchy like a phenomenological laser. But the freewheeling *chora* with its insurgent energies and multiple significations offers new possibilities for dealing with masculinist erasure. *From this place of nonidentity, ecofeminists boldly reframe the nature—woman—labor nexus, revaluing what has been problematic in a one-dimensional order so as to confront its stagnant totalization.*[27] Some liberal feminists and even some socialists, still speaking from the unreconstructed side of the Woman/Nature contradiction, fail to see the dialectical shift here, and so they call ecofeminist thought "essentialist." This is not surprising, since the scientific hegemony of capital cannot handle irony, the moment of tension when a signifier is suspended between two competing senses. Further, the power of bourgeois realism is such that the very term "essence" itself is captured by positivism, losing its negative, unmasking function.

Far from the complacent certainties of positivism, the negative dialectic holds to an inverse relation between power and historical consciousness. However, it is not the free-floating liberal intellectual who has privileged access to the critical perception; nor is a theory of class consciousness adequate to understand the material contradiction that positions women and nature together against men of capital and labor. Rather, ecofeminist insights are usually driven by the profound "lack" imposed on those who are neither "human" nor of "nature." In Adorno's words, nonidentity is "the somatic unrest that makes knowledge move;" the dialectician's duty is to help this "fool's truth" attain its own reasons.[28]

6. Politics and Sustainability

By reasoning dialectically, then, ecofeminists introduce an alternative ontology to political discourse, one that cancels the frightened dualisms produced by masculinist denial of woman and nature. Ecofeminists propose that:

- Nature and history are a material unity.
- Nature, women, and men are at once active subjects and passive objects.
- The woman–nature metabolism holds the key to historical progress.
- Reproductive labors guided by care are valuable models for sustainability.

Tying political perception and motivation to suffering, the phenomenology of deconstruction that women experience results in a materially grounded "epistemology from below." Concerned with equality for all life forms, ecofeminism is a *socialism* in the very deepest sense of that word. And it may be noted that "spiritual" ecofeminism reflects these same ontological assumptions. The contribution of this feminine voice is even more apposite to *ecology*, as men begin to regard nature itself as a subject with its own needs. Both dominated and empowered, women are well equipped at this conjuncture to take up the case for "other" living beings. Again, this is not to argue in a simpleminded essentialist way that women are somehow "closer to nature"—as patriarchal ideology would have it, in order to keep women in their place. Rather, it is to acknowledge a complex socially elaborated sex and gender difference, privileging women temporarily as historical agents par excellence.

The most urgent and fundamental political task is to dismantle ideological attitudes that have severed men's sense of embeddedness in nature; and this, in turn, can only happen once nature is no longer fixated, commodified as an object, outside of and separate from humans. Reifications of this latter sort are endemic to capitalist discourse, starting with the very subject of "bourgeois right" who is supposed to participate in the democratic process with a fixed identity and status. Socialism too has traditionally attributed a permanent character to the proletariat as historical agent. But universals or essences like "humanity," "class," "woman," and "nature," are static abstractions that do violence to those living under the regime of contradiction. The alternative ecofeminist conception of subjectivity as signification-in-process, permanently forming and reforming itself in collision with the social order is based in a living and embodied materialism that defies the limits of bourgeois epistemology.

Against the theoretically abbreviated notions of capitalist patriarchy, the ecofeminist consciousness is reflexively decentered. Walking the primary contradiction of capital, women activists must engage in a zigzag dialectical course between (1) their liberal feminist task of establishing the right of women to a political voice; (2) their radical and

socialist feminist task of undermining the very basis of that same valida-
tion by dismantling the capitalist–patriarchal relation of man to nature;
and (3) their ecofeminist task of demonstrating how women—and
thence men too in future—can live differently with nature.[29]

Unlike capitalist partriarchy, which is geared to short-term
profits, women's lives straddling the *nature–woman–labor nexus* are
embedded in a context of conservation and care. Transcending the
limits of both capital and socialist ideologies,

> if women's lived experience were . . . given legitimation in our
> culture, it could provide an immediate "living" social basis for the
> alternative consciousness which [radical men are] trying to formulate
> as an abstract ethical construct.[30]

Thanks to capital and its contradictions, ordinary women, a
global majority, already model sustainability in their cycle of reproductive labors. The labor of Finnish housewives described by Pietila,[31]
or of Indian women farmers, instantiates this. Here, in practice, are
ways of meeting community needs with low disruption to the environment and minimum reliance on a dehumanizing cash economy.
Honoring the "gift" of nature, such women labor with an independence, dignity, and grace that people looking for sustainable models
can learn from. For, as Shiva reminds us,

> Culturally perceived poverty need not be real material poverty:
> subsistence economies which satisfy basic needs through self provi-
> sioning are not poor in the sense of being deprived . . . millets are
> nutritionally far superior to processed foods, houses built with local
> materials are . . . better adapted to the local climate.[32]

Unlike women's work, the market economy is disconnected
from daily physical realities, its operational imperatives bear no relation to people's needs; its exponential "growth" trajectory even kills
off its own future options as it goes. As global capital becomes increasingly centralized by the transnational management of information
flows, nation-states are rendered powerless and working men made
marginal in a labor force segmented by practices like enterprise bar-
gaining and subconstracting. The situation of women, as housewives
in "advanced" industralized societies, regresses to a point where they
no longer control either their means of production or their own
fertility. Their domestic maintenance functions continue to echo the
"mediation of nature" for men, but these women lose skills and
autonomy to consumerism, while the very manufacture of so-called
labor-saving "products" destroys the living habitat beyond repair.

Ecofeminists reject the idea that "necessary labor" is a burden to be passed on to nature through technology. Equally, they reject a strategy of "partnership" with the union movement in an unviable economy. Maria Mies calls for a notion of labor as pleasure and challenge.[33] And most ecofeminists look forward to self-sufficient, decentralized relations of production, where men and women work together in joy and reciprocity with external nature, no longer alienated or diminished by a gendered division of labor and international accumulation. Ecofeminism is about a transvaluation of values; in particular, it is about listening differently to the voices of women who love and labor now.

Notes

1. The reified and essentialist constructions "nature," "woman," "labor," are used intentionally, representing the discursive materiality that has to be resisted in political struggle. Even so, both the universalizing patriarchal term "woman" and its empirical form "women" are social constructions. Some postmodern writers query the category "women" as well, in an attempt to avoid "essentialism." Marxists, with a universalizing concept of "class," and people of color who write about "race" risk the same charge. However, a structural analysis of domination(s) cannot be made without recourse to general categories. The present essay does not differentiate women by stratifications of class, race, age, and so forth, since the *nature–woman–labor nexus* crosses these conceptual boundaries. This is one reason for arguing that it constitutes a primary contradiction.

2. This approach in terms of a critical embodied materialism is sketched out in Ariel Salleh, "On the Dialectics of Signifying Practice," *Thesis Eleven*, 5–6, 1982, pp. 72–84. Compare Sandra Harding, *The Science Question in Feminism* (Ithaca, N.Y.: Cornell University Press, 1986), for another articulation of standpoint epistemology.

3. Nancy White, quoted by Patricia Hill Collins, *Black Feminist Thought* (Cambridge, Mass.: Unwin, 1990), p. 160.

4. International Labor Organization (ILO) statistics adopted by the U.N. in 1980. The original wage percentage estimated by the ILO was 5%. Subsequent U.N. publications raised the figure to 10%.

5. The notion "conditions of production" as used by James O'Connor in his "Capitalism, Nature, Socialism: A Theoretical Introduction," *CNS*, *1(1)*, no. 1, Fall 1988, brackets together physical resources, human labor, and local infrastructure. Here I suggest that these categories beg more differentiated analaysis to account for the *nature–woman–labor nexus* as a primary contradiction of capitalism.

6. See, for example, Carole Pateman, *The Sexual Contract* (Cambridge, U.K.: Polity Press, 1988).

7. Selma James, *The Global Kitchen* (London: Housewives in Dialogue Archive, 1985), p. 1.

8. Marilyn Waring, *Counting for Nothing* (Sydney, Australia: Allen and Unwin, 1988).

9. Mary Mellor, *Breaking the Boundaries* (London: Virago, 1992).

10. James, *The Global Kitchen*, op. cit., pp. 10–11.

11. James, *The Global Kitchen*, op. cit.; Waring, *Counting for Nothing*, op. cit.; Hilkka Pietila, "Women as an Alternative Culture Here and Now," *Development*, 4, 1984; Australian Bureau of Statistics figures for 1990, quoted in the National Women's Consultative Council Report, *A Question of Balance* (Canberra: Government Printer, 1992), Appendix F.

12. See, for example, Lydia Sargent, ed., *Women and Revolution* (Boston: Southend Press, 1981).

13. Berit As, "A Five-Dimensional Model for Social Change," *Women's Studies International Quarterly*, 4, 1981.

14. See Irene Dankelman and Joan Davidson, eds., *Women and Environment in the Third World* (London: Earthscan, 1988); Lin Nelson, "Feminists Turn to Workplace, Environmental Health," *Women and Global Corporations*, 7, 1986; and Cynthia Enloe, *Bananas, Bases and Beaches* (London: Pandora, 1989).

15. Vandana Shiva, *Staying Alive: Women, Development and Ecology* (London: Zed, 1989).

16. Ibid., p. 45.

17. See, for example, E. A. J. Johnson, *Predecessors of Adam Smith: The Growth of British Economic Thought* (New York: Prentice-Hall, 1937), pp. 139–140.

18. John Locke, "An Essay Concerning the True Original, Extent and End of Civil Government" (circa 1688), in Sir Ernest Barker, ed., *The Social Contract: Essays by Locke, Hume, Rousseau* (Oxford: Oxford University Press, 1971), especially part V, paragraphs 25–51, pp. 16–30. The citations are from p. 17 and p. 18, respectively; emphasis in original.

19. Naomi Scheman, "Individualism and the Objects of Psychology," in S. Harding and M. Hintikka, eds., *Discovering Reality* (Boston: Reidel, 1983), p. 234.

20. Maria Mies, *Patriarchy and Accumulation on a World Scale* (London: Zed, 1986).

21. The need for "self-criticism" in this regard is put forward by Italian socialists Valentino Parlato and Giovanna Ricoveri, in a recent presentation on "The Second Contradiction in the Italian Experience," *CNS*, 4(4), no. 16, December 1993.

22. Paul Eluard, "Sound Justice" (1951), in Gilbert Bowen, ed. and trans., *Paul Eluard: Selected Poems* (London: Calder, 1987), p. 145.

23. Ariel Salleh, "Contribution to the Critique of Political Epistemology," *Thesis Eleven*, 8, 1984.

24. Giovanna Ricoveri, "Culture of the Left and Green Culture: The Challenge of the Environmental Revolution in Italy," *CNS*, 4(3), no. 15, September 1993, p. 119.

25. Julia Kristeva, *Polylogue* (Paris: Editions du Seuil, 1978). *Cathexis* is a psychoanalytic term implying an investment of nervous energy.

26. See Ariel Salleh, "On the Dialectics of Signifying Practice," op. cit. This usage of the "negative dialectic," with its *negative (nonpredicated)* "essence," comes from Theo Adorno, *Negative Dialectics*, E. Ashton, trans. (London: Routledge, 1973), and *Minima Moralia*, E. Jephcott, trans. (London: New Left Books, 1969).

27. In a rather more rationalist vein, Collins also writes about the outsider/within experience of black women as a positive epistemological stimulus. "Black women's lives are a series of negotiations that aim to reconcile the contradictions separating our own internally defined images of self as African-American women with our objectification as the Other"; *Black Feminist Thought*, op. cit., p. 94. The international activist group Development Alternatives with Women for a New Era similarly advocates a feminist standpoint epistemology based on the view from below. A critique of postmodernist objections to this approach is contained in my forthcoming book on ecofeminism (London: Zed).

28. Adorno, *Negative Dialectics*, op. cit., p. 203; *Minima Moralia*, op. cit., p. 73.

29. See Ariel Salleh, "The Ecofeminism/Deep Ecology Debate: A Reply to Patriarchal Reason," *Environmental Ethics*, *14*, 1992, for a more detailed exposition.

30. Ariel Salleh, "Deeper than Deep Ecology," *Environmental Ethics*, *6*, 1984, p. 340.

31. Pietila, "Women as an Alternative Culture," op. cit.

32. Shiva, *Staying Alive*, op. cit., p. 10.

33. Mies, *Patriarchy and Accumulation*, op. cit.

7

On the Misadventures of Capitalist Nature

Martin O'Connor

The social control of air, water, etc., in the name of
environmental protection evidently shows men entering
the field of social control a little more deeply themselves.
That nature, air, water become rare goods entering the field
of value after having been simply productive forces, shows
men themselves entering a little more deeply into the field
of political economy. At the limit of this evolution, after
natural parks, there may be an "International Foundation
of Man" just as in Brazil there is a "National Indian
Foundation": The National Indian Foundation is in a
position to assure the preservation of the indigenous
population in the best possible conditions, as well as [sic]
the survival of the animal and vegetable species that have
lived alongside them for thousands of years." (Of course
this institution disguises and sanctions genocide and
massacre: one liquidates and reconstitutes—same schema.)
Man no longer even confronts his environment: he himself
is virtually part of the environment to be protected.[1]

1. Introduction

Environmental crisis has given liberal capitalist society a new lease
on life. Now, through purporting to take in hand the saving of the
environment, capitalism invents a new legitimation for itself: the

Lecturer in Economics, University of Auckland, New Zealand. With special thanks
to Giovanna Ricoveri, Danny Faber, and other members of the Boston CNS group,
to Ariel Salleh, and to James O'Connor. First published in Italian, in *Capitalismo Natura
Socialismo (CNS–Italia) Anno Terzo*, no. 8, pp. 45–79; English version published in
CNS, 4(4), no. 16, December 1993, pp. 7–40; slightly abridged for this volume.
Some of the themes were laid out in my earlier sketch, "The System of Capitalized
Nature," *CNS, 3(3)*, no. 11, September 1992, pp. 94–99.

sustainable and rational use of nature. The purpose of this essay is to provide an exposition and critique of this hegemonic escapade.

In particular my argument will focus, in theoretical terms, on the process of the *capitalization of nature* as a response within capitalism to (1) the ostensible *supply problem* of depletion of natural resources and degradation of environmental services required for support of commodity production, and (2) the resistance by communities and whole societies to the ecological and cultural depredations wreaked by expanding capital. By "capitalization," I mean the *representation* of the biophysical milieu (nature) and of nonindustrialized economies and the human domestic sphere (human nature) as reservoirs of "capital," and the codification of these stocks as property tradeable "in the marketplace"—saleable at a price that signifies the value (utility) of the goods and services flows as inputs to commodity production and in consumption.

My central argument will be that, through the capitalization of nature, the modus operandi of capital as an abstract system undergoes a logical mutation. What formerly was treated as an *external* and exploitable domain is now redefined as itself a stock of capital. Correspondingly, the primary dynamic of capitalism changes form, from accumulation and growth feeding on an external domain, to ostensible self-management and conservation of the *system of capitalized nature* closed back on itself.

This process of what we can call the *semiotic expansion of capital* is aided by the coopting of individuals and social movements in the "conservation" game. But the outcome is not harmony and conservation; rather, it is a terrible, abject competitiveness on all counts—a political and military struggle to have particular interests and capitals valorized at the expense of others, and to lay claim to scarce resources (raw materials and environmental services) needed to assure sustenance of the particular interests or capital stock. In the resolution of this struggle, there is no such thing as a meaningful "balancing of interests,"[2] but rather the *fact and imminence* of annihilation for the less favored interests and beings who are the "used" and abused, and the *fear* on the part of the "users" of losing their privileged place and eventually becoming themselves crushed by someone else's production designs or left to wither along the wayside.

The degraded production conditions and nonavailability of required services that appear, abstractly, as a threat to the accumulation of someone's capital, are also, materially, "the means of life, means of survival and consumption, and, in the case of laborpower, life itself."[3] So, while capital has a crisis concerning the costs/feasibility of its own reproduction and accumulation, people have a crisis concerning the viability and regeneration of their material and social conditions of life. Correspondingly, while the contradictions inherent in the capitalization process can be analyzed abstractly (as I do here) as contradictions at the political and economic levels of the capitalist accumulation-reproduction process, behind this abstraction are the personal, existential, and social dimensions of struggle and crisis.[4] So this theoretical mode of analysis has to be seen as simply one of the many dimensions of human *resistance* to capitalist accumulation and to the capitalization process.

2. From Cost Shifting to the Capitalization of Nature

In traditional Marxist analysis crisis is the occasion that capital attempts to seize and to turn to its own ends, that is, to restructure and rationalize itself in order to restore its capacity to exploit labor and nature, and thereby accumulate surplus value to itself. Capital responds to environmental crisis with, in the first instance, attempts to extend its hegemony over the sources of needed raw materials and services for commodity production. James O'Connor has suggested that when faced with an environmental supply-side crisis, capital will attempt to *restructure* the conditions in order to reduce costs. Many individual capitalist enterprises, hell-bent on survival in a competitive world, would prefer to continue to treat nature and societies as open-access terrains to be mined and trampled upon at will. This is illustrated not only by military or quasi-military operations of force majeure, but also by the great pressures on governments, for example, not to "lock up" lands against mining, not to capitulate to "unbalanced" demands of consumer and community interest groups seeking preservation of nature reserves and wilderness against development, or not to impose stringent pollution controls and obligations to safeguard workers' health.

But capitalistic enterprise, through the cumulative effects of this *cost shifting*, tends to destroy the conditions of production on which it depends. This is the "second contradiction of capitalism" as formulated by James O'Connor and others: attempts by individual capitalist enterprises to maintain or improve profitability through free-riding on existing conditions of production (physical environment, labor, and social services and infrastructures) lead to higher costs of production for capital as a whole.[5]

Moreover, such attempts at maintaining favorable (to capital) cost and supply conditions for needed raw materials and services (of nature, of labor, of society as socialization force and infrastructure) can involve fairly obvious dispossession, as well as cost shifting onto local communities, onto "the taxpayer," and onto future generations. Politicization of environmental issues is a clear sign of social resistance to this process.[6] Where simple appropriation and cost shifting are impossible, an alternative (and in many ways more farsighted) tactic of relegitimation is the *capitalization* of the conditions of production. This means designating as valuable stocks erstwhile "uncapitalized" aspects of the physical environment (nature) and of civil society (infrastructure, households, and human nature). From the point of view of capital, the delineation of clear "property rights" over natural domains facilitates their highest value use. At the same time, local communities and social movements (and, more generally, societies and peoples) may be enticed to cooperate, through representing them as the *stewards of the social and natural "capitals"* whose sustainable management is, henceforth, both their responsibility and the business of the world economy.

People are rightly skeptical about the "recognition" thereby afforded to heretofore external elements. More often than not, this capitalization is not a signal of authentic respect and sustenance, but works instead as a vehicle for disappropriation, dispossession, and continued cost shifting on a huge scale. What we see unfolding is a terrible charade in which capital expands its hegemony through actually *feeding off* the resistance from the elements of human nature which, along with their habitats, are being liquidated and commodified (rationalized and reshaped, as in corporate takeovers). Capital makes the *management* of these liquidations and reconstitutions into a new source of dynamism. The goals of protest movements are travestied, and conservationist rhetoric becomes an unheralded boon in capital's own project of enlarged reproduction.

3. The System of "Capitalized Nature"

Formally, the capitalization process involves bringing a new element or set of elements within the commodity domain through a *process of colonization*. Elmar Altvater, writing in *CNS*,[7] has made the point that

> the expansionist pressure inherent in the economic logic of surplus production has a territorial dimension (as production is necessarily always spatial). Surplus production is thus identical to economic conquest—exploration, development, penetration, and exploitation—of space, i.e., "production of space." At first, space is conquered extensively; subsequently it is capitalized intensively.

He went on to say that this "propagandist tendency of the world market" leaves nothing untouched on the entire globe, and he is right. He also makes clear that this penetration, which involves in the first instance an invasion, plunder, and despoliation, is equally a *semiotic conquest of the territory*. This involves a sort of double play around the distinction capital/nature. Capital proclaims as rational and proper the appropriation of *nature* as "free," as desired input materials and services. But, then, if brute appropriation is contested by the groups concerned, the stratagem of capitalization is used to ensure, and to legitimate, access—still, as we will see, at the lowest cost practicable. In this motion, as James O'Connor has highlighted, abstract capital effects the *ideological production* of "capitalist nature," this term denoting "everything which is *not* produced as a commodity but which is treated *as if a commodity*."[8] In application to the physical environment, this usually takes the form of creating marketable property rights, for example, over forests, fisheries, or water sources. By this ideological movement, the erstwhile exterior domain (of "nature") is redefined as an element of valuable capital, present within the world productive system, and itself to be *rationally managed* as a productive enterprise.

Through this process of *internalization of production conditions*, we are seeing the emergence of an "enlarged" system of capital that differs in some fundamental ways from the capitalist systems of the 19th century. We may speak thus of a "mutation in the system of capital." The hypothesis of such a mutation is not new. Quite apart from being foreshadowed in a number of works of speculative fiction, it was a thesis systematically developed by Jean Baudrillard in a number of essays during the 1970s.[9] In a 1972 essay, translated as "Design and Environment, or How Political Economy Escalates into Cyberblitz,"[10] he described this mutation whimsically as the implementation of "the doctrine of participation and of public relations extended to all of nature":

> Nature (which seems to become hostile, wishing by pollution to avenge its exploitation) must be made to participate. With nature, at the same time as with the urban world, it is necessary to recreate communication [i.e., to institute harmony] by means of a multitude of signs (as it must be recreated between employers and employees, between the governing and the governed, by the strength of media and of planning). In short, it must be offered an industrial contract: protection and security—incorporating its natural energies, which [will otherwise] become dangerous, in order to regulate them better.[11]

Here Baudrillard makes a parallel with the treatment of human labor in Western economies. Through the 19th century the wages paid to industrial labor permitted bare survival in squalid conditions. Neither industrialists nor governments paid much attention to the conditions required for regeneration of the labor force. The domestic/communal domain of human reproductive activity, and of regeneration of labor power more generally, was placed on the side of "nature." Relative to capital, it was an *external domain*. Recognition of the "productive" value of the household/domestic/subsistence sphere was limited to the value of the wage paid to those household members participating in commodity production. The household/domestic sector was, at the same time, required to absorb the extra burdens of unhealthy working conditions: congestion, pollution, and so on—costs related to the activities of capital itself.[12]

Then, as is well known, the mid-20th century saw the fantastic augmentation in the West of commodity consumption, through mass consumerism. Responding to the "demand-side" crisis of the 1920s and 1930s, a combination of Henry Ford's "high wages = high sales" formula and Keynesian pump priming worked to shore up Western capitalist production systems right through the 30 glorious years after World War II. On the ideological plane, the induction of people into mass consumption functioned as capital's alibi. Even if denounced by the likes of Marcuse as involving creation of "false needs" by the system, and by more traditional Marxists on account of the exploitation of labor power it effects, the provision of bountiful "use values" to the mass of citizens has been successfully represented as the triumph of consumer sovereignty.[13] As Baudrillard has put it,[14] the workers are liberated as consumers, but this is only in order to better serve the expanded reproduction of capital. This coopting of the household comes to be perfected, in ideology, through the formal theory of *human capital*, which preaches that the household as domain of consumption and source of labor power is a *business enterprise* having its own objectives and its own capital to be managed. The money earnings received by a household are, correspondingly, the returns to this human/domestic capital. So, from being simply an external domain of exploitation, tapped into as needed by capital, the household and community is now represented as a *productive sector* that manages its own inputs, and in which valuable capital can be invested in anticipation of a profitable return.

A similar analysis can be made of the coopting of non-Occidental societies into the world economy. Native (non-Westernized) human beings were viewed by Western colonists, ideologues, and political philosophers as undeveloped human nature, while the territories of "primitive," nomadic, and peasant societies were viewed as empty and undeveloped nature—treated alternately as dangerous wildernesses, wastelands, and virgin un-broken-in territories. As such, the lands, waters, and societies were plundered for raw materials (agricultural surpluses, minerals, fish, slaves, women), and made into a domain of sport and adventurism (for everything from big game safaris to sexual favors).[15] Then, through the processes of "development" and "modernization" has come a wholesale transformation of these societies: the destruction of preexisting social relations, living spaces, myths, and physical milieux, and their reconstitution or replacement by forms serviceable to capital. Formerly places of autonomous creative regeneration, these terrains and their peoples come to be represented as passive sites where the Western entrepreneurs come to inject their capital, supply their technicians (usually male), and direct the production of commodities of value.[16] Here *the market* is called upon to do its job. Formally, the individuals and societies in question are acknowledged as the rightful proprietors of (what is left of) their own capitals. These are valuable stocks: genetic, material (maternal), labor, cultural. These "natural" domains will, henceforth, be treated as valuable capitals: slave labor is replaced by wage labor; plunder is replaced by mining concessions.

The important theoretical point is that, through this process of *capitalization* of all domains of raw materials and services, and through the *internalization* via the extension of the price system considered as "susceptible to giving account of everything and to directing all processes,"[17] capital undergoes a qualitative change in form. No longer does it simply exploit better and more intensively a nature (and human nature) external to itself. In what we might call the *ecological phase of capital*, the relevant image is no longer of man acting on nature to "produce" value, henceforth appropriated by the capitalist class. Rather, the image is of nature (and human nature) codified as *capital incarnate*, regenerating itself through time by controlled regimes of investment around the globe, all integrated in a "rational calculus of production and exchange"[18] through the miracle of a price system extending across space and time.

This is nature conceived in the image of capital; and this representation of nature is the basis for the "rational management" of nature/capital that, increasingly, is *instituted violently* in political fact.

4. Sustainability and Capital's Conservation

This characterization and analysis of a global *system of capitalist nature* involves a high level of abstraction, whose purpose is to help sharpen analysis of observed events. The distinction being drawn between the two "forms" of capital is foremost a logical one. In actual fact, capitalist enterprises continue to operate in a predatory fashion on uncapitalized domains of nature and humanity; and these external domains, although preyed upon, remain largely uncontrolled by capital itself. Predation and cost shifting go hand in hand with the rhetorics of environmental preservation and heritage conservation. So we have primitive exploitative accumulation, on the one hand, and the rhetoric of "sustainable management" of the *system of capitalized nature,* on the other hand. Which description is more appropriate? Often both; but here it is a matter of understanding the logic of *representation* and the instituted forms that the power struggles take. Within the mature system of capitalized nature, the struggles for survival take on the appearance of a reciprocal (but uneven) *autophagy of capitals,* where the proprietor of a dominant capital is making illicit (or at any rate contested) use of the capital of another, if not using it up altogether.

The questions we must ask in looking at capitalist expansion are: What is expanding? What is growing? What (if anything) is being sustained? Through the capitalization process, the total stock of capital is augmented by the addition of imputed values for virgin forests, fish stocks, gene pools, mineral deposits, land considered as a recreational amenity, and so on. But this augmentation of values has no strict correlation with an expansion of physical activity as such. Rather, capital's expansion is a sort of semiotic outgrowth, as in a tumor, that progressively envelops its surroundings.

The colonization by the European powers of the New Worlds, and the linking of the South to the industrialized powers through development of commodity markets and trade, have involved forcible expropriation of human and material energies that might otherwise have found a place in reproduction of the nonindustrial economies, and have imposed various material and symbolic burdens to the point of poisoning the peoples and asphyxiating the local cultures. Relative to this expropriation and suffocation, it might be supposed that recognition of "value" and of "ownership" through the capitalization process is a step toward justice, affording belated protection for and respect of the individuals and societies not just as raw materials, means to an end, but as autonomous enterprises in their own right. The proclaimed objective of the 1992 Rio de Janeiro Earth Summit was to *save the planet:* to save natural heritage, cultural heritage, genetic diversity, vernacular life-styles, and so on. After all, if capital is nature and nature is capital, the terms become virtually interchangeable; one is in every respect concerned with *the reproduction of capital, which is synonymous with saving nature.* The planet as a whole is our capital, *which must be sustainably managed.*

Sadly, though, this rhetorical harmonization does not in any way guarantee the conservation of specified productive or reproductive potentialities of a society or ecosystem, nor does it assure the sustaining of the particular interests, communities, or ecologies thus valorized. In practice, the main effect of all the identifying of "at-risk" heritages, stocks, and capitals is the better and better alignment—in ideology, though not necessarily in fact—of this participant nature (and human nature) to the norms of capital's own enlargement and reproduction.

Recall the notion central to traditional Marxism that the "finality" of capital is essentially its own survival and expanded reproduction as a system of social control. It is always the capitalist *system* that is to be reproduced and sustained, not any individual capital. We may apply this idea to capital ecologized, but with a radically altered meaning. In the mature system of capitalized nature, every raw material, every input to production from a terrestrial source, is formally recognized as being an element of capital or as a service derived from a capital. As such, it has an owner, and its use means an allocation of capital or of services that may be derived from a particular capital—for which the owner should be paid. Every *useful* output is one with the potential to satisfy a need of the system—which means, one way or another, being in the service of reproduction of another capital. Whether this "use" be represented as consumption or as production is immaterial: it is in either case an *investment in the service of reproduction of the system of capital.*

So the "bottom line" in capital ecologized as a whole is not accumulation as such. Certainly, individual firms and corporations will (of necessity) adhere to the imperative of profitability, for those that fail are out of the game. The owner of a capital will be paid if, and only if, he or she puts it to the service of other capital; and unless paid, will not be able to command inputs of needed materials. But what matters for *the system of capitalized nature* as a whole is no individual capital or capitalist, but rather the sustaining over time—the *conservation*—of the system itself as an abstract social form.[19]

To unravel this perversion of the conservation rhetoric, it is useful to consider separately several different features of the workings of capitalism and the requirements of conservation. Section 5 presents in a stylized way the requirements for simultaneous conservation of all valorized capitals in an interdependent ensemble. Section 6 then presents the reasons why, even if such a conservation goal were deemed desirable, its rigorous achievement is probably impossible, and certainly unlikely under capitalist norms. Section 7 goes on to show, with some examples, the way in which selective extension of property rights to natural domains effects dispossession rather than respect or protection of the former proprietors and beneficiaries of the domains or resources in question. Finally, Section 8 discusses ways in which capitalism coopts the efforts of local communities and protest groups by encouraging the mentality of "valuing"—in dollar terms, and in a decontextualized manner—the various social and environmental spillover burdens of capitalist enterprise.

5. Sustainability and Solidarity

Consider, abstractly, the requirements for sustainability in the sense of simultaneous maintenance of the stock levels of a set of interdependent "capitals" making up the system of capitalized nature. Suppose we have a system of interdependent "*sectors*," such as manufacturing, households, forests, and so on, each one with a characteristic "*capital*" and each sector providing services or produced surpluses to other sectors, and receiving inputs from others. It is convenient to assume that every capital has a human proprietor (a society or social group), and that this human proprietor is sustained along with the capital(s) they "own."

Suppose, now, that all the capital stocks are to be *conserved* as viable going concerns. We will say that the ensemble of capitals is *sustainable* in a material sense, as long as sectors give to each other, and receive from each other, exactly what each needs for self-maintenance. This would mean that each sector gives to the other sectors (as a group) what the latter need so as to furnish in return the inputs required by the former. For simplicity of argument we can assume zero net growth of the capital stocks, meaning all surpluses would be used up in the process of ensemble reproduction.[20]

Suppose further that each capital's (i.e., each sector's) proprietors must pay for inputs received, and that they obtain payment for services and surpluses rendered to others, and that this is done according to a set of prices. The ability to *pay for*—and by this ability to *command* (if physically available)—the required inputs depends on income received for surpluses and services rendered. In a zero-growth steady-state pattern of activity, revenues received by each sector will exactly balance expenditures made for each "production period."[21]

Now consider various ways in which there might be an "imbalance" in the supply–demand situation as it affects a particular sector. First, with regard to inputs for sector maintenance, a "supply crisis" can arise if the required inputs are physically not available or if the price is higher than the sector proprietors can afford, given the revenues they are receiving for their own services/surpluses rendered. The consequence of an enduring supply crisis of this sort is that the sector goes without adequate inputs, meaning *starvation* and hence the diminution or cessation of activity. In the case of physical penury of inputs, the only hope is substitution of a different source of nourishment. If the problem is high prices for inputs, this might be accommodated by selling off capital(s)—which means *depletion of stocks*, again violating the conservation criterion being assumed. Moreover, if one sector withers away or is depleted, then it will not be able to furnish inputs required by other sectors. So there will be "transmission" of supply crises by domino-effect from one sector to another.

Second, with regard to outputs of services/surpluses from a particular sector, a "demand-side" crisis may arise if there is "no market" for what is furnished. This may occur if no other sector has a "use" for the outputs in question[22] or if the sectors having a use for them cannot afford to pay for them. In either case, the producing sector faces a "realization crisis." An enduring demand-side crisis of this sort means a drop in revenues received, which will curtail the sector's own ability to purchase the inputs it needs for its own sustenance, thus meaning a price-induced supply-side crisis and this may in turn lead to a demand-side crisis transmitted to other sectors.

Locally, a crisis may be experienced due either to physical unavailability or to inability to pay. In terms of our conventions, supply takes place only when a money payment is made (or promised). We may think of the "demand" for a material or service as an indication of willingness to pay, which entices a valuable gift from one sector to another. If at any stage a sector finds itself unable to make a valorized gift (i.e., sell its surpluses/services for a "good" price), it will become—under these conventions—unable to entice desirable gifts to itself. Either the sector will collapse from neglect (inadequate nourishment) or the proprietors will have to sell off some capital; in either case, this can amount to the loss of the basis for well-being or survival of the proprietor group themselves.

This stylized analysis enables us to identify two basic requirements for simultaneous conservation of all capitals. First, all the sectors must, as a set, be *potentially compatible* with each other. This means that there exists the possibility of a pattern of intersectoral transfers ("trades" in a broad sense) that ensures maintenance of all capitals through time. Second, such a pattern must be *actually maintained*, meaning that the sectors must operate in solidarity with each other rather than with some accumulating surpluses at the expense of others (through predation and/or leaving others to starve to death), or some being self-maintaining while dumping their toxic wastes and unwanted refuse on others, and so on.

6. Capitalist Accumulation and "Nature's Resistance"

These ideas of solidarity, predatory accumulation, and realization crisis are, of course, familiar in traditional Marxist (as well as Keynesian) literatures. Here, though, we are concerned with the question of simultaneous sustainability of "natural capitals" in addition to industrial capital, whereas traditional Marxism followed liberal political economy in treating the "natural" domains as *external to capital* and exogenously determined. This changes the whole tenor of the management problematic.

Technocrats of all political persuasions like to dream that the "metabolism" of nature (and human nature) can be rendered fully predictable and controlled, as with the regulation of industrial commodity production and exchange processes. In the back of the minds of both those advocating centrally planned "solutions" and those supporting market "solutions" to environmental problems is the idea of *economic equilibrium* for a set of commodity production processes and intersectoral exchanges. They imagine that, for the extended system of capitalized nature, one could find a set of "shadow prices" that would signal the correct exchange proportions for the services and material resources—the exchange values which, if instituted, would establish an "equilibrium" or "sustainable balance" in the system's operation.

However, in the case of nonhuman capital, we are dealing with domains (ecosystems, land, atmosphere, seas, etc.) where human ignorance is high, and which, while sometimes "managed" by human occupants, are neither wholly initiated by humans nor controllable in the same sense as manufacturing processes.[23] From an instrumental reason standpoint, this refractoriness can be comprehended only negatively. Nuclear reactor malfunctions, chemical spills, and runaway genetic experiments are contemporary paradigms of the *accident ecologized*, where malfunction in one part of the system ramifies throughout. The problem of the accident, however, emerges directly as a dialectical by-product of the Enlightenment ideology of mastery of nature by technology. It is a sign not just of the nonrealization, but more particularly of the *nonrealizability* of perfect functional control. As Mexican writer Octavio Paz puts it,

> This frailty of things does not have its existence in reality as such: it is a property of the system, something which pertains specifically to the logic of the system. The Accident is neither an exception nor an illness of our political regimes, nor is it a correctable defect of our civilization: it is the natural consequence of our science, of our political and moral tradition. The Accident forms a part of our idea of Progress.[24]

The more we advance in technological prowess, the deeper the resultant perturbations to ecological, atmospheric, metabolic, and psychological processes, and the more imponderable the "side effects" of this progress in production.

Moreover, it cannot be taken for granted that the "natural" and the "industrial" sectors will have a symbiotic potential. While human economies are dependent on "natural capital" sectors for inputs and for waste-receipt services, the major biosphere cycles do not have the same dependency on human agricultural or industrial processes. Industrial processes in particular may be antagonistic to many ecological processes, either because of the scale of their raw materials needs, or because of the disruptive effects of their "waste" outputs, including many materials that are inimical to human and nonhuman life forms. This puts in doubt the feasibility, *even in principle*, of long-run simultaneous conservation of the full spectrum of existing ecological capitals alongside existing types of industrial capital.

It needs to be clearly understood, then, that this project of total capitalization and "sustainable management" of the global system operates primarily at the ideological, or *social imaginary*, level; and correspondingly, it is a largely *imaginary* management that is instituted. The image of "controlled participation of nature" in sustainable development is a vicious fraud. Equally, it can by no means be presupposed that the reproduction or "development" aspirations of households, communities, and whole societies, formerly external to capital but now assimilated to it as "human capital" sectors, are compatible with current modes of industrial growth and accumulation. Under its mask of the rational use of resources, the enlarged system of capitalized nature is in permanent ferment, and acutely conflict-ridden. World capitalism is characterized not only by incessant corporate dogfights, but also by struggles between capitalist interests and social groups resisting dispossession of their own lands, resources, and selves, alongside conflicts within and between these noncapitalist social groups for possession of resources and over priorities in conservation, use, and transformation of ecosystems. The refusal of people to submit to capital's designs (notwithstanding the pacifications effected through industrial workplace discipline and consumerism, the stupefactions of TV, etc.) is both a proof and a cause of the lack of full control of capital over human nature.

7. The Dynamics of Dispossession

The myth of liberal society, of the marketplace as a just institution, is that the self-interested pursuit of profits—capital accumulation—can, through the miracle of market exchange, be a win-win game. The proposition is that people can relate to each other exclusively as *means to their own ends* (furnishing inputs to production/consumption) while still respecting them as ends in themselves (consumers/proprietors of capital). This mythology willfully ignores the full dimensions of human and ecological interdependencies (euphemistically describable as the problem of externality), and the real disparities in access to resources in a materially finite world.

What would it mean for all individuals and societies to be able to enjoy radical liberty as producers and consumers, that is, freedom of choice as self-determining owners and proprietors of capital? In brief, this would require, at the very least, (1) that there exist some pool of resources freely available to all individuals and societies, that each may draw upon "at liberty"; and (2) that the ambient environment be able to absorb the repercussions of these proprietors' diverse uses of these resources—in commodity production, liberation of energy, disposal of wastes, and so on—such that each person's enjoyment does not impinge on others' persons and property.

What happens, then, if we postulate that all biophysical territories are bespoken, meaning the finiteness and *scarcity* of natural capitals—resources and sources of environmental amenity—relative to demands? This is what has become the issue today, as property claims extend over deserts and Antarctic lands (for oil and minerals), the global atmosphere, the high seas and seabeds—and through time. How should these valuable capitals be apportioned between peoples and across time if, as the rhetoric goes, all individuals and entire cultures are allowed the "right" to determine their own path of "development"? We cannot seriously suppose that, in a finite world, all the putative rights, resource demands, and requisites for waste disposal are simultaneously realizable. No. If we define the criteria of sustainable development like this, our only conclusion could be that *someone's or something's* interests must be inhibited, trampled, imposed upon. The supposed exercise of liberty at this point becomes inseparable from *doing violence*, that is, from its own antithesis.

Of course, violence in the sense of domination is the leitmotif of the capitalist project. For those pursuing capitalist accumulation, supply-side crisis takes on meaning only when resource extraction, the environmental side effects of production, or resistance by affected social groups reach sufficient crisis proportions to impair the availability of raw materials and services sought by capital's proprietors themselves. If ready substitutes for used-up materials, labor, environmental services, and sites can be found, or if a shift to different commodity lines not requiring the same inputs can profitably be made, supply crises are readily resolved. Only when political opposition is overwhelming or substitution is not possible does the imperative arise that the environmental sites/sources be *managed sustainably and conserved.*

Manifestly, such self-interest in profits does not equate with an authentic interest in these sources as life forms or social ends in themselves! The dominant responses of capitalism to environmental crisis and to demands for respect of cultural difference continue to be premised on an instrumental, if not downright cynical, treatment of nature and human nature. There continues to be direct appropriation of supposedly "free" natural domains, which in general means exclusion of other human groups.[25] Even when the accumulation activity is admitted to involve some use of or despoliation of human property, a variety of devices are used to obtain what is wanted at the lowest price to capital, without regard to whether the price paid or the manner of use are compatible with long-term reproduction requirements of the capital in question. Moral accountability is filtered through the veil of the "price system," with the premise that one's obligations to the other party are acquitted by the "agreed" price for a good/service rendered.

For example, in New Zealand, legal and political battles are currently being fought by the indigenous Maori people for practical control of, and recognition of, their "ownership" of large tracts of lands, shorelines, and coastal waters traditionally in tribal use as food sources. Coastal and deep-sea fishing had, in large part, remained "open to the public" until the 1970s, with widespread small-scale activity for local food-gathering and recreational purposes. By the 1970s, however, a rapid rise in levels of commercial fishing—mostly for export, with much of the deep-water fishing done by offshore companies—was threatening depletion of many fish species. The 1980s saw the capitalization of the resource, through introduction of a tradeable quota regime limiting total annual catches for at-risk species. This system was designed to ensure the sustainability of the *commercial* fishing industry! The scheme as initially implemented curtailed catch rights for many individuals and entire communities, many of them

Maori, active for decades in "noncommercial" fishing—a flagrant dispossession and a real threat to the communities' livelihoods.

The legally instituted dispossession of local interests in favor of corporate capital was effected through the Crown[26] pretending that originally "no one owned the resource"; so it was, arguably, the government's to take in hand. The 1980s management procedure was, first, to commodify access to the fish species most under threat, in the form of catch "quota"; and then to award these rights to the major commercial operators as a free gift, pro rata according to their documented catch histories. The small-scale and "informal" operators, and the local people who thought that they enjoyed an environmental domain as a collective heritage and source of sustenance, were told that they did not "own" it at all. Effectively, ownership (all commercial catch rights) were awarded to the large commercial operators who, from the Maori/local points of view, were the original "poachers" overfishing the stock.

Most recently, as more threatened species are being brought under the quota system, one Crown proposal has been to have potential users tender for the catch rights. To those currently fishing legally (under open access for the nonquota species), the question now being put to them is: "What are you willing to pay in order to keep your rights to catch fish?" This proposal has understandably met with widespread opposition. For small-scale operators within local communities this feels a bit like, in a family home, having the eldest son pick up the family cutlery and announce: "I can get $5,000 for this in the market. Are you prepared to offer me more for it?" A group's "willingness to pay" is constrained by disposable income, and almost invariably large commercial operators harvesting the resource for export and profits can pay more hard cash for a given amount of quota than can a one-boat operator, local family group, or even a whole Maori tribal group, whose "valuation" of their catch is not necessarily translated into cash sales or tradeable assets backed by banks.

Of course, in awarding catch rights, the structure of privilege might have been put the other way around, with ownership of the "capitals" in question being placed in the hands of local or indigenous claimants. The question posed to local interests, say, Maori tribes, then would be: "What must a commercial user pay you as minimum compensation (lowest acceptable price) in order that you will yield the asset up to the marketplace?" The award of capital ownership to a group with a conservation interest (rather than a short-term profit interest) might seem more favorable to conservation than dispossession under the willingness-to-pay rule. But it still does not guarantee that wherewithal for the capital's renewal—or for the social groups' sustenance—will be maintained through time. On the contrary, given

the real political dominance of capital, such award of capital ownership can in practice mean the coopting of the local/tribal interests to
the interests of capital, through representing them as stewards of
various capitals—of themselves as human capital, and then also environmental, community, cultural assets. From this point on, the market makes an incessant enticement to the local tribe to "profit" from
its asset—to sell not just fish caught but the entire capitalized catch
rights. If the proprietor group is self-sufficient and has little desire
for the commodity riches obtainable through the marketplace, and/
or if it has an enduring interest in the maintenance of the capital (e.g.,
fish for the grandchildren), then long-term maintenance of the group
and their capital is a plausible outcome. But if the group needs or
want to buy commodities in the market, it may feel pressure to exploit
the cash-generating potential of the capital, either through the sale of
a stream of fish caught, or through lease/sale of the capital (the catch
rights) itself. As soon as the catch or leasing is made under commercial
pressures, or the rights are sold outright to a commercial operator,
the risk increases of unsustainable levels for short-term cash gain,
undermining long-term conservation interests. Moreover, even internationally recognized legal ownership does not necessarily furnish
effective protection against cost shifting or predation by other agencies. Self-sufficient and conservation-minded proprietors may not be
able to defend themselves against damage from external sources, such
as commercial and recreational poachers, contamination by industrial
wastes, a marine accident that releases oil or plutonium, or other
forms of ecological degradation.[27]

What this example makes clear is the systematic bias in the way
that "property" or "capital" is defined—in favor of appropriation by
commodity-producing capital. The logic of the marketplace states
plainly that all capitals will realize their "full value" only by their
insertion within the sphere of exchange value. Under the doctrine of
utility maximization, their best use will be signaled by price: they
should always go to the highest bidder in the marketplace. But as
Baudrillard puts it in a convenient formula, "Under the pretext of
producing maximal utility, the process of political economy general-
izes the system of exchange value."[28] It is the *market system* (under
the alibis of capital accumulation and increasing consumption) that is
to be sustained.

8. A Price for Everything?

Similar mechanisms of dispossession are at work in the processes of
"taking into account" social and environmental costs imposed by

commodity production on local communities round the world. If we think of each aspect of the physical environment as a capital stock (natural/built, as the case may be) with defined proprietors/beneficiaries, then any action by one group or individual that impairs the flows of benefits—present and future—to these proprietors/beneficiaries, may be considered as a *cost shifted* onto the latter. In usual economic jargon, a cost or burden imposed without a compensating payment made is a *negative externality*.

Negative external effects in this sense are not rare or exceptional. They are the tangible signs of interdependency in a finite world. They are the dialectical complements of the exchanges "in the market" which are so cutely idealized as purposeful, controlled, and contractual. They are an integral part of modern life, just as much as accidents and natural disasters are the other side of the Enlightenment coin of instrumental control. In reality, unreconciled reproduction and accumulation projects accommodate themselves unhappily to each other, meaning that some prevail, and some don't. The struggle for survival—political, military, guerrilla resistance, legal-economic contests—takes the form of a battle for control of scarce capitals and the benefits deriving from them, to ensure viability of favored reproduction or accumulation projects. The upshot is *reciprocal (though uneven) predation and cost shifting* among the various proprietors and claimants of capitals.

Corporate capital will seek to subordinate and coopt the available labor, social infrastructure, and environment to its own profit. But also, social movements learn how to enunciate and defend their own conservation (henceforth: natural, cultural capital reproduction) projects against predation and cost shifting. In any given case, the firms' profit or the movements' conservation goals may or may not be achieved. But for the capitalist *system*, it does not matter where the burdens ultimately fall, as long as the game itself continues to be played. In the end, as Baudrillard says, through the pricing, capitalization, and liquidation processes, "the system becomes the only bearer of active functionality, which it redistributes to its elements."[29]

In this era of generalized scarcity, *everything has its price*, and the degraded environment itself is the new commodity to be carved up, managed, bought and sold. If an environmental amenity such as a wetland or a neighborhood park is under threat from, say, a shopping mall or a motorway, then prima facie there might be a case for compensation to be paid. But how should the value of the loss be assessed? Who pays and who should be paid?

If, after legal–political contest, the amenity is deemed the property of "the community," then the latter should—under libertarian precepts—be given full compensation for their loss. The "development" interests will thus find their costs escalated or their supply lines blocked. An analogous situation occurs when a sovereign government withdraws concessionary privileges to transnational mining, fishing, oil-extracting, or timber-milling interests, or imposes severe conditions relating to pollution control or waste disposal. So the enterprise may become unprofitable. Alternatively, it might be deemed that no one "owns" the environment. Then every interested party is asked, What price would you be willing to pay? We can then compare the value placed on the amenity by community interests with what a productive enterprise would be willing to pay for the use of the site, materials, and services. How do future generations make their bids?

Whatever variation of this valuation procedure is employed, one ensures—supposedly—that each unit of capital will go to its highest valued use. Once this principle of "best use" has been established, it can be admitted that, of course, there will be some unrequited desires and some losers. It will be pointed out, though, that the projects that missed out represented less highly valued uses of the resources, hence were suboptimal in the greater scheme of things.

Let us look, in this light, at the question identified by James O'Connor of "the rising costs of reproducing the conditions of production."[30] Such varied costs as counseling, divorce fees, and medical care to repair damage and stress suffered by the modern work force, crime control, and environmental restoration after mining, building, or industrial accidents, are, he points out, "all unproductive expenses from the standpoint of self-expanding capital"; and presently in the United States, "total revenues allocated to protecting or restoring production conditions may amount to one-half or more of the total social product."

It would be an error to suppose that these expenditures, large though they are, indicate the "full costs" of capitalist destruction. In the first place, not all damages are put right. Full restoration is in many cases impossible, and compensation in money terms is not always possible. The damages are often *irreversible*, and if there is loss of life or if no adequate substitute can be made or found for a central constituent of a social group's life (material, symbolic), the value of the losses may properly be judged infinite. So in practice "compensation" and "repair" will necessarily be piecemeal and incomplete. There are real losers. Who are the winners? James O'Connor qualifies the social and environmental repair costs as unproductive *from the standpoint of self-expanding capital.* Yet it must be emphasized that the modus operandi of modern capital in its ecological phase is not profit as such but a semiotic domination. What matters is to *institute socially* the commodity form, so as to represent all of nature (and human nature) as capitals, ipso facto in the service of capitalism as a legitimate social form. Looked at overall, the "pricing" of a good (or of a bad), or the successful capitalization of an element of nature, or the successful repulsing of a shifted cost, signals a semiotic conquest, namely, the insertion of the elements and effects in question within the *dominant representation* of the overall capitalist system activity. Unproductive as expenditures on health, urban transport systems, historic building repair, environmental cleanup, waste disposal, and so on, may be from the standpoint of individual firms' profits—and even *overall* global profits and accumulation—nonetheless, all these extra costs to be priced, and all these reclamations of *values to be taken into account and conserved*, can become good currency for modern capital. They have an undoubted "use value" for the project of reproduction of capital as a *form of social relations.*

Recall that, for industrial capitalism, it does not matter at all if a few (or a few hundred) firms go bust, as long as the down-valuation of their assets can be represented as a sign of progress—for example, a sign of technical advance or of higher productivity elsewhere, or a consequence of changing consumer tastes. So if certain enterprises go to the wall because they must internalize some social costs, or because less favorable conditions of supply mean an upward spiraling of their input costs, this does not matter for capitalism *as a system* so long as the *remaining* players in the game have identified for themselves a capital to be managed and reproduced, a commodity or service to be marketed, priced, and sold.[31]

For capitalism ecologized, similarly, what does it matter if a previously ignored environmental "value" surfaces, compelled into being by the imminent threat of its negation, or by the negation of other similar values; it is nonetheless profitable (to some) so long as it can be *re-presented* as a capital, profitably to be conserved and used. What does it matter if some genes, Indians, tropical forests, urban environments and their inhabitants, and so on, are, along with the firms forced to the wall, lost in the rush. These losses are of little consequence as long as the outcomes and decisions can be *re-presented* as the *rational use* of the available capital: that is, at some moment in time, by somebody's calculus, on whatever shifting-sands evaluation baseline, allocated to their highest value on the margin use.

The overriding *political* task faced by capital is to stave off its latent bankruptcy as a modality of social organization, of rationality. The keynote in this task is that the populace be put to work articulating the "values" that capitalism itself is placing at risk. Political legitimation of capital depends on getting people to believe in the capitalization process as a defense against the predations of capital! The *discourse of "taking everything into account"* does not ensure real protection or repair. Overall there is no "net gain," indeed, there is no consistent unit of measure by which such an assessment might be made. Having a price only signifies the quasi-functionality, the "use value" of each resource or object *within (and to) the system of capitalized nature*. What alone counts is the successful generalization of the code of exchange value as a *semiotic operation*. This is a question that transcends profit-and-loss accounting. It is, as Baudrillard puts it,[32] "a matter of life and death" for the survival of capitalism as a social system: "Perish profits, in order that the reproduction of the form of social relations be preserved."

Crisis-ridden capitalism manages, through the generalization of the price system, to make the *management* of the liquidations and (partial) reconstitutions a strategic terrain in pursuing its own project of sustainable development. What gets lost in the fury of the contest is the hegemonic power of cost-benefit valuation discourses as a vehicle for legitimation of capital *even as the liquidations take place*.

9. Concluding Remarks: Resistance and Reciprocity

The indeterminacy inherent in the current struggles over the purposes to which environmental resources will be put can be given a prophetic ring, as, for example, by James O'Connor locating us historically in a period,

> in which nothing can be taken for granted, in which the mobilization of resources and good political strategy and leadership will decide in one way or another whether the conditions of production are defined as desired by capital, or by labor, communities, and urban populations.[33]

At the same time, the coiling discourses of communities' and indigenous peoples' rights, and of environmental values to be taken into account, must be seen to have a tragic side. Individually, when a community group wins a battle, obtains compensation for health damage from a toxic spill, or from a dangerous drug, or saves a mountainside from being strip-mined, we rejoice. But it would be disastrous if people were to believe that taking account of the "true value" of things (and of people) had anything at all to do with doing a full and correct cost–benefit analysis.

The cost–benefit valuation logic is internal to and derivative of the instrumental, egotistical, competitive logic of capitalism. Its use tends to reproduce contradictions inherent in the capitalist process, without resolving them. Under the prevailing situation of world political domination by capital, the main effect of the pricing-capitalization processes is to signify all the items are being valued as *means to the end of capital accumulation*. This is most explicit when the "willingness-to-pay" rule is applied, as this presumes nonownership a priori, and leaves all potential users—capitalists and environmentalists alike—abject before the arbitrary power of the marketplace. Yet even when ownership of environmental capitals by "local interest" groupings is formally recognized, the effect is a mobilization of the resources—their entry into the sphere of exchange value—in the larger interests of capitalism as a dominant social form.

In dog-eat-dog market society, winners "use" *and use up* losers—and the "losers" include individuals qua workers and consumers (and qua "unemployed" nonworkers and nonconsumers), and collectivities qua firms, communities, and indeed whole societies, ecosystems, and species. In the mad scramble for survival, the sheer plethora of values articulated and the incoherence of all these fragmentary valuation efforts works as a smokescreen deflecting attention from the outright impossibility, even in theory, of a "rational management" of all this notionally capitalized nature. For example, many *relatively* exploited peoples in countries of the North are fighting legitimately for a minimum of health and for economic survival, yet are locked into defending a "life-style" that is inherently exploitative of the economies of the South and disdainful of these peoples' livelihoods.

So the question still remains: The (sustainable?) management of production conditions, as desired by *which* labor, *which* communities, *which* urban (and rural) populations? While one can be sure about the *relative injustices* in many situations, this discourse of general grassroots mobilization still leaves us far from understanding the basis for, the possibilities and limits of, *the real choices involved in* a humane coexistence in a finite world.

Abstractly, this problem of coexistence probably defies satisfactory solution. In practice, solutions have to be, more or less unhappily, *worked out*. A radical resistance to the depredations of capitalism will have to be based on a refusal to embrace the logic of the system of capital itself—that is, *repudiation of the idea and resistance to the real process of commodification/capitalization* of nature, of labor, and of infrastructure. Correspondingly, a respectful coexistence of diverse life forms and "capitals"—human and otherwise—would have to be pursued on the basis of *relinquishing* (1) the Enlightenment control myth, and (2) the norm of self-interested accumulation. So-called rights are only one half of the story, the other being a willingness to accept the obligations (material and symbolic) and limitations of reciprocity. These are the terrains to be theorized, and explored in practice, on global and local scales in coming years.

Notes

1. Jean Baudrillard, "Design and Environment, or How Political Economy Escalates into Cyberblitz," in *For a Critique of the Political Economy of the Sign* (St. Louis: Telos Press, 1981), English translation by Charles Levin of *Pour une critique de l'économie politique du signe* (Paris: Gallimard, 1972), pp. 185–205.

2. In fact, terms such as "maintaining a balance," "taking everything into account," "ecological balance," and so on, are *inherently duplicitous* and are best not used at all. It is always a question of which life potentialities will prevail, and which not; which interests prevail, and which not. On this point, see, for example, R. Arnoux, R. Dawson, and M. O'Connor, "The Logics of Death and Sacrifice in the Resource Management Reforms of Aotearoa/New Zealand," *Journal of Economic Issues*, 27, December 1993, pp. 1059–1096.

3. James O'Connor, "The Second Contradiction of Capitalism: Causes and Consequences," in *Conference Papers, CES/CNS Pamphlet Series no. 1* (Santa Cruz, Calif.: Center for Ecological Socialism, 1990).

4. This point is made particularly plain by analysts writing from materialist ecofeminist perspectives in relation to women's labor, land, water, and environmental degradation, for example, Maria Mies, *Women: The Last Colony* (London: Zed Books, 1988); Vandana Shiva, *Staying Alive: Women, Ecology and Development* (London: Zed Books, 1989); Brinda Rao, "Women and Water in Rural Maharashtra," *CNS, 1(2)*, no. 2, June 1989, pp. 65–82; Ariel Salleh, "Living with Nature: Reciprocity or Control?," in J. R. Engel and J. G. Engel, eds., *Ethics of Environment and Development* (London: Belhaven Press/Pinter, 1990), pp. 245–253, and "Class, Race, and Gender Discourse in the Ecofeminist/Deep Ecology Debate," *Environmental Ethics*, 15, 1993, pp. 225–244.

5. James O'Connor, "Capitalism, Nature, Socialism: A Theoretical Introduction," *CNS, 1(1)*, no. 1, Fall 1988, pp. 11–38; and "The Second Contradiction of Capitalism," op. cit. See also various contributions to the "Symposium on the Second Contradiction" in *CNS*, various issues.

6. A number of writers have analyzed environmental movements as social responses to actual or threatened "externalities" of the world market system. These movements "perform a function at which the market fails," by signaling the presence of costs that the market ignores, and by requiring that to some degree these burdens be mitigated or payment made for them. On these themes, see, for example, Juan Martínez Alier, "Ecological Eco-

nomics and Ecosocialism," in *CNS*, *1(2)*, no. 2, June 1989, pp. 109–122, reprinted in this volume as Chapter 1; and "Distributional Obstacles to International Policy: The Failures at Rio and Prospects after Rio," *Environmental Values*, 2, 1993, pp. 97–124.

7. Elmar Altvater, "Ecological and Economic Modalities of Time and Space," *CNS*, no. 3, November 1989, pp. 59–71; revised and reprinted in this volume as Chapter 4.

8. James O'Connor, "Capitalism, Nature, Socialism," op. cit., pp. 7, 23.

9. See notably Baudrillard's *Pour une critique de l'économie politique du signe*, op. cit.; *Le Miroir de la production* (Paris: Casterman, 1973), English translation by Mark Poster: *The Mirror of Production* (St. Louis: Telos Press, 1975); and *L'Échange symbolique et la mort* (Paris: Gallimard, 1976), especially part I entitled "La Fin de la production" (The end of production).

10. In Baudrillard, *Political Economy of the Sign*, English trans., op. cit., pp. 185–203; henceforth cited as *Cyberblitz*; all page references will be to the English translation.

11. Baudrillard, *Cyberblitz*, p. 201.

12. Indeed, the very real work carried out—substantially by women, as numerous feminist writers have pointed out—in reproductive, food gathering and processing, home maintenance, communal support, and sexual "services," has been largely ignored by most "liberal" theorists of industrial capitalism. See, for example, Mary Mellor, "Eco-Feminism and Eco-Socialism: Dilemmas of Essentialism and Materialism," *CNS*, *3(2)*, no. 10, June 1992, pp. 51–52, recapitulating arguments made, in a variety of ways, by many socialist feminist and ecofeminist writers, including Susan Griffin, Maria Mies, Ariel Salleh, and Marilyn Waring. Indian activist and philosopher Vandana Shiva asserts that, under Western development ideologies, women's labor has in effect been "pronounced to be nonlabour, to be biology; their labourpower—their ability to work—appears as a natural resource and their products as akin to a natural deposit"; see "The Seed and the Earth: Technology and the Colonization of Regeneration," unpublished manuscript, 1992.

13. And this despite the fact that, for many people, this new "consuming" involved much new work: consumption of fossil fuels and transport services commuting to and from work or school, household chores maintaining and upgrading the increasingly gadget-filled household. For example, the roles ascribed to the "household" in waste disposal and recycling of packaging junk in Germany today is nicely analyzed by Irmgard Schultz, "Women and Waste," *CNS*, *4(2)*, no. 14, June 1993, pp. 51–63.

14. See especially Baudrillard, *The Mirror of Production*, op. cit.

15. Since, of course, the natives could prove dangerous, in colonizing practices there is no clear divide between hunting wild animals for sport, conquest of women, and the subordination of an unruly alien human nature. . . .

16. Vandana Shiva, "Seed and the Earth," op. cit., p. 12, gives a graphic image from the sphere of medicalized human capital reproduction, which

can serve as a metaphor: "When women are having children, they are viewed less as sources of human regeneration than as the 'raw material' from which the 'product,' the baby is extracted. In these circumstances the physician rather than the mother comes to be seen as having produced the baby." Of course, in civilized societies, the mother receives in "exchange" for her various domestic labors (housework, reproductive and sexual services) diverse considerations in cash and in kind (which may include domestic purposes welfare benefits for solo mums, and surrogate motherhood fees); but these revenues accrued are not necessarily enough to keep in good repair the mother's *own* body and soul.

17. Baudrillard, *Cyberblitz*, p. 192.

18. Ibid., p. 188.

19. Baudrillard, in *L'Échange symbolique et la mort*, argues that this *conservative role* has always been at the heart of the capitalist project. He thus comments: "What is reproduced in the present-day system is capital in its most rigorous definition: as a *form of social relations*, and not in its vulgar acceptation as money, profit and economic system. Reproduction has always been taken to mean the 'enlarged' reproduction of the mode of production, and as such determined by this latter. But one should, rather, conceive of the mode of production as a modality (and not the only one) of the *mode of reproduction*. Productive forces and relations of production—in other words, the sphere of material productivity—are, perhaps, only one of the possible conjunctures, therefore historically relative, of the process of reproduction. The category of reproduction is a form that goes far beyond economic exploitation alone. The play of productive forces is, therefore, not the necessary condition of it" (p. 49; my translation). This leads to the question, What exactly drives capitalism, in symbolic and cultural terms? Baudrillard suggests that behind the preoccupations with endless accumulation, with power, and with holding stock, is a fear and refusal of death (being felt as annihilation, negation of the ego). Fear of the void, fear of "the Other"; whereas accumulation is more of the same (an image of immortality, empty for every human, but achievable, by projection, through capital's reproduction. . . .).

20. The essentials of the arguments to follow do not change if positive growth (capital accumulation) is allowed, but rigorous exposition is messier. Thinking of capitals in material terms, however, the zero-growth assumption concords intuitively with the idea of *conservation* of interdependent economies and ecosystems comprising the mass-closed planet earth. Formally, it also gives us a benchmark which means "unbalanced" growth corresponds to predatory activity: expansion of one sector's capital through, directly or indirectly, depletion of another. All the arguments to follow can be formalized with the aid of input-output algebra; but this is hardly necessary here.

21. If formalized in algebraic terms, this "balance" result is an analogue of well-known accounting balances in intersectoral analyses of economic systems.

22. Many by-products of industrial production would fall into this "unwanted" category, being useless or sterile (if not toxic and destructive) from the standpoint of other economic and ecological reproduction activities. I

discuss this theme within a thermodynamic and model framework in "Entropy, Liberty and Catastrophe: The Physics and Metaphysics of Waste Disposal," in Peter Burley and John Foster, eds., *Entropy in Economics* (Dordrecht: Kluwer, in press).

23. See my discussions in Martin O'Connor, "Codependency and Indeterminacy: A Critique of the Theory of Production," *CNS*, *1(3)* no. 3, November 1989, pp. 33–57, reprinted in this volume as Chapter 3; and in "Entropy, Liberty and Catastrophe," op. cit.

24. Paz, *Conjonctions et Disjonctions*, as cited by Baudrillard, *L'Échange symbolique et la mort*, op. cit., p. 246; my English translation.

25. The dispossession and annihilation of the remaining Amazonian "Indian" tribes along with the jungles themselves is part of our modern-day folklore. Yet this is simply a continuation of centuries of plunder: refer, for example, to Eduardo Galeano, *Las venas abiertas de América Latina* (Mexico City: Siglo XXI Editores, 1971), English translation by Cedric Belfrage: *Open Veins of Latin America: Five Centuries of the Pillage of a Continent* (New York: Monthly Review Press, 1973). In practice, the recognition of the "equal rights" to development and self-determination of non-Western peoples seems, as Serge Latouche puts it, to be a recognition post mortem; see Latouche, *L'Occidentalisation du monde: Essai sur la signification, la portée et les limites de l'uniformisation planétaire* (Paris: La Découverte, 1989), p. 138.

26. That is, the state of government, as New Zealand's "postcolonial" structure of governance is a constitutional monarchy within the British Commonwealth.

27. This is a manifestation of the general problem of spillover impacts and predatory behavior in a competitive world. Now that the New Zealand fishing "quota" rights are capitalized, this actually amounts to a battle over "assets" valued in hundreds of millions of dollars. In the quest for dollars, side effects on other marine life or ecosystems are often ignored. One might well ask about "intrinsic values" of marine ecosystems, and whether the fish have standing as life forms in their own right? (And then: Who speaks for the fish and sea mammals, and with whose money in their wallet?) For further discussions of these issues, see Martin O'Connor, "Valuing Fish in Aotearoa: The Treaty, the Market, and the Intrinsic Value of the Trout," and Leith Duncan, "The ITQ System: A Critical Appraisal," both in *Justice and the Environment: Proceedings of the 17 June 1993 Symposium*, published in the *Policy Discussion Papers* (Department of Economics, University of Auckland, February 1994); and also Richard M. Dawson, "The Fisheries Claims Act and John R. Commons' Reasonable Value," *Policy Discussion Paper*, no. 14 (Department of Economics, University of Auckland, February 1993).

28. Baudrillard, *Cyberblitz*, p. 191.

29. Baudrillard, *Cyberblitz*, p. 197.

30. James O'Connor, "Capitalism, Nature, Socialism," op. cit., p. 26. See also Frank Beckenbach, "Social Costs in Modern Capitalism," *CNS*, *1(3)* no. 3, November 1989, pp. 72–92, reprinted in this volume as Chapter 5. These reproduction/maintenance costs constitute what, elsewhere, Ivan Illich has called the counterproductivities of modern production.

31. A similar analysis can be made concerning the enormous debt carried by Third World countries. What matters, for the hegemonic power of capital, is that the debt be "managed sustainably." It does not matter if it is ever paid off (in fact, this is almost certainly impossible, and anyway, capital is more powerful if it is not!). Nor does it matter if, in the various gyrations, restructurings, and reschedulings, numerous banks go bust, nor if some of the burden is written down against conservation covenants for tropical forests (as in Brazil), and so on. The pass-the-parcel game of debt management becomes a sort of mutual hostage-taking on a world scale, binding everyone into the game. See, on these points, Jean Baudrillard, *Les Stratégies fatales* (Paris: Grasset, 1983), English translation: *Fatal Strategies* (New York: Semio-text(e), 1990), pp. 34–49: "Neither dead nor alive, the hostage is suspended by an incalculable outcome. It is not his destiny that awaits him, but anonymous chance, which can only seem to him something absolutely arbitrary." Also see Baudrillard, *La Transparence du mal* (Paris: Galilée, 1990), pp. 33–42, on "Transéconomique": "The extraordinary vision of this debt which rotates, of these missing capitals which circulate, this negative wealth which will, undoubtedly, one day be listed on the Stock Exchange along with everything else" (my translation).

32. Baudrillard, *L'Échange symbolique et la mort*, op. cit., p. 43; my translation.

33. James O'Connor, "The Second Contradiction of Capitalism," op. cit., p. 5.

8

Is Sustainable Capitalism Possible?

James O'Connor

1. Introduction

There are few expressions as ambiguous as "sustainable capitalism" and such sister concepts as "sustainable agriculture," "sustainable energy and resource use," and "sustainable development." This ambiguity runs through all of the most important discourses on economy and the environment today—U.N. and government reports, scholarly research, popular journalism, and green political thinking. Precisely this obscurity leads so many people so much of the time to talk and write about "sustainability": the word can be used to mean almost anything one wants it to mean, which is part of its appeal.

"Sustainable capitalism" has both a practical and a moral ring to it. Who in his or her right mind would be against "sustainability"? The earliest meaning of *sustain* is "support," "uphold the course of," or "keep into being." What corporate chief, treasury minister, or international civil servant would not embrace this meaning as his or her own? Another meaning is "to provide with food and drink, or the necessities of life." What underpaid urban worker or landless peasant would not accept this meaning? Still another definition is "to endure without giving way or yielding." What small farmer or entrepreneur

Martin O'Connor's editorial help is warmly appreciated. This chapter is a much revised version of "Is Sustainable Capitalism Possible?," which appeared in Patricia Allen, ed., *Food for the Future: Conditions and Contradictions of Sustainability* (New York: John Wiley and Sons, 1983), pp. 125–137, and a lecture, "Discourse on Sustainability," given at the School for Environmental Studies, York University, Ontario, Canada, March 1993.

does not resist "yielding" to the expansionary impulses of big capital and the state, and thereby take pride in "enduring"? There is a struggle, worldwide, to determine how "sustainable development" or "sustainable capitalism" will be defined and used in the discourse on the wealth of nations. This means that "sustainability," in the first place, is an ideological and political, not an ecological and economic, question.

In the present account, the word *sustain* is taken to apply to all three of the above senses: to "uphold the course" of capitalist accumulation globally; to "provide the necessities of life" for peoples of the world; and to "endure without yielding" by those those ways of life are being subverted by the wage and commodity forms. The question of sustainable capitalism thus pertains in part to whether or not sustainability defined in all three ways can be achieved, and how it can be achieved.

There is a fourth meaning of *sustain*, namely, "ecological sustainability," even though there is little agreement among ecological scientists about the exact meaning of this expression. For example, "biodiversity" or "planetary health" are rarely problematized in terms of ecological science and the ideologies embedded in this science, nor is the expression "ecological crisis," which is widely used by popular writers without benefit of a clear definition. One definition of the latter might be "a turning point during which it is decided whether a species, ecosystem, bioregion, or the planet as a whole lives or dies." Yet "ecological crisis" defined in this way has no status within ecological science (nor could it, given that ecology is an ambiguous combination of atomistic and holistic assumptions about the laws governing living nature). Population ecologists and conservation biologists normally correlate population changes of a particular species; changes in "carrying capacity" defined narrowly in terms of the needs of that species; and some coefficient that measures the relationship between the species and the carrying capacity in question, on the one hand, and the rest of the ecosystem, which the species in question may depend on in indirect ways, on the other. All these terms have some explanatory power. But this multiplicity of determinants means that there is no obvious way to really know whether or not threats to an individual species are self-inflicted, so to speak, or arise because of changes in the ecosystem as a whole. If this is so, talk about the "sustainability" of particular species may be less precise, and the concept of "environment crisis" more problematic, than would superficially appear to be true.

These ambiguities become even more pronounced when ecologists or greens mix social and economic with biophysical dimensions, and discuss the "sustainability" of whole ecosystems or bioregions. In California's Monterey Bay region, for example, excessive pumping has lowered the water table, causing salinization from sea water, threatening the viability of agriculture. Is this a "crisis"? In economic terms, not if the region imports water; in fact, imported water might breathe new life into local agriculture, as well as housing, commercial, and industrial development. "Sustainable agriculture" means one thing if a strict bioregional perspective is adopted and something else if the perspective is widened to include other bioregions. In this particular case, it turns out that the debate over importing water has less to do with the "sustainability" of local agricultural capital and water quality, and more with normative judgments pertaining to what kind of community and culture people in the region want to have (i.e., in Pajaro Valley, whether to keep its present Mexican cultural flavor or to open the area more to commuter populations from Silicon Valley on the other side of the coastal range).

Defining *sustain* in these four ways, the short answer to the question "Is sustainable capitalism possible?" is "No," while the longer answer is "Probably not." Capitalism is self-destructing and in crisis; the world economy makes more people hungry, poor, and miserable every day; the masses of peasants and workers cannot be expected to endure the crisis indefinitely; and nature, however "ecological sustainability" is defined, is under attack everywhere.

In this chapter I will review some important evidence bearing on the problem of "sustainable capitalism," highlighting along the way the very different concepts of "sustainability" deployed by greens and corporations. A brief account of the conditions of economic sustainability (or of profitability and accumulation), narrowly defined, is offered. I will then discuss the "internal" or "first" contradiction of capitalism, and the crisis–ridden and crisis–dependent nature of capitalist accumulation, appending a short review of the gathering world crisis in the 1980s. The argument is made that the prospects of global economic management are as dim as those of global environmental regulation. Next, I discuss another seemingly intractable problem (a "second" contradiction) facing capitalism today, namely, a "cost-side" profit squeeze generated by the contradiction between capital and nature (and other conditions of production), together with the adverse economic effects on capital of environmental and other social movements. The ways in which capital is trying to confront these issues is discussed. I discount capital's capacity to successfully deal with both the first and the second contradiction, thanks to the nature of the liberal democratic state and of capital itself. I then underline the very uncertain political—hence economic and ecological—consequences of a general economic depression. Finally, following a short review of environmental conditions in the poor countries (the South), I draw some conclusions about the possibilities of radical environmental or "red green" social and political movements. While the prospects for some kind of "ecological socialism" are not bright (so the argument goes), those of a "sustainable capitalism" are even more remote.

2. Environmental Policy and the Discourse on Sustainability

The evidence favors the judgment that capitalism is not ecologically sustainable, despite the recent flood of talk about "green products," "green consumption," "selective forestry," "low-input agriculture," and so on. In the 1992 U.S. presidential campaign not one of the three major candidates made the "environment" an important issue. The Reagan and Bush governments compromised themselves on issues ranging from the use of federal grazing land to logging old-growth forests to fighting pollution (abandoning tried and true methods of pollution control for "market solutions"). Increasingly, state and local governments neglect the environment in their competition to attract scarce capital. In federal law, the definition of "wetlands" was narrowed, as was that of "endangered species." Occupational health and safety enforcement measures established over decades were undermined. National and state parks are now more commodified, as managers search for ways to meet expenses. Oppositional movements are fragmented. True, the Clinton administration is showing a "green" color in reforming the use of public grazing lands; nuclear power is temporarily stalled; some capital good industries such as paper and pulp have begun to install cleaner technology; and organic farming has benefited from a surge of consumer interest in pesticide-free products. But the majority of union leaders oppose or are indifferent to most demands made by environmentalists, and established environmental organizations (with two or three notable exceptions) are more willing to compromise their positions in the name of "economic growth."

In most countries green parties remain small or are accommodating their demands in national and local politics. In Europe, the environment is not a central concern of the bureaucrats who run the powerful European Commission, despite representation by greens in the European Parliament. International agreements on ozone layer depletion are weak and those on global warming are merely symbolic.

Agreements with respect to protecting the world's "commons"—watersheds, forests, rivers, lakes, coastlines, oceans, and air quality—are more often than not honored in the breach. Whaling may be revived and fisherpeople everywhere clamor to empty the waters of their bounty. Oil as an instrument of economic wealth and national power is more important than ever before. Energy and mining companies (often the same) are poised to massively exploit more mineral resources everywhere from upstate Wisconsin to Siberia. In the South, many governments are eager to sell their natural birthrights to transnational corporations, often under the pressure of big external debts, in the name of "development," and the landless and land-poor masses of the world's countrysides and the urban poor are forced to deplete and exhaust resources and pollute water and air respectively, simply to survive. The environmental records of the East Asian "tigers," the Southeast Asian "little tigers," and Mexico, Brazil, and other Latin American growth centers are not encouraging.

One necessary step, practically speaking, toward ensuring a sustainable capitalism—defined in some sense of "ecologically rational or sound"—would be national budgets that put high taxes on raw material inputs (e.g., coal, oil, nitrogen) and certain outputs (e.g., gasoline, chemical building blocks), meanwhile slapping value-added taxes on a wide range of environmentally unfriendly consumer products (cars, plastic products, throwaway cans)—complete with an enforceable "green label" policy that would exempt genuinely green products, with "green" defined strictly in terms of ecological impacts at every stage of the production, distribution, and consumption process. Other steps would include national expenditure policies that heavily subsidize solar energy and other benign alternative energy sources; technological research that leads to eliminating toxic chemicals and other substances at their source; innovations in mass transit; improvements in occupational health and safety conditions coupled with national, regional, and community enforcement procedures; and a redefinition and reorientation of scientific and technological priorities generally. In few political entities is this kind of green budget—with appropriate changes in methods of national income accounting—being developed, except on paper by a marginalized group of green economists and activists.

At the level of the discourse on "sustainability," the prospects for an ecologically sound capitalism, recognizable to greens, seem problematic at best. In fact, behind a seeming convergence of vocabulary is a disjuncture or gap between green and capitalist discourse, with both sides talking past each other. One problem is the discourse of much of the environmental movement, supported by companies that seek to green themselves, or at least present a green image to the public. This discourse seeks to find ways for corporations to reform their economic practices to make them consistent with the sustainability of biodiverse forests, water quality, wildlife preservation, atmospheric conditions, and so on. Focus is placed on production processes, technology, recycling and reuse, and energy efficiency, as well as broader questions pertaining to the structure of consumption, finance, marketing, and corporate organization, and also government policies. For example, the reform-minded World Resources Institute recently stated that sustainability presupposes an "unprecedented transformation" of technology. For reformist greens, then, the question is how to remake capital in ways consistent with the sustainability of nature.

In the boardrooms of many corporations, however, the problem is discussed in different terms. At a superficial level, the issue is simply one of how to present a plausible green image to consumers and the public (e.g., the U.S. chemical industry planned to spend $10 million in 1992 to paint itself as environmentally reasonable and friendly).[1] It is also a question of how to introduce practices into production that save energy and raw materials. However, this has a largely economic motive. Far from being a problem for capital as a whole, energy and material efficiency in a period of slow growth (or a world recession that threatens profits) is economically desirable. To take one instance, as much as 75% of aluminum produced by U.S. companies today is made from recycled aluminum cans and other products. New wood industry practices that make posts and beams from trees too small to cut up for lumber, thereby utilizing what in the past would have been discarded as waste, is another example. Also, "recycling" rhetoric and (selective) practices can be used to facilitate new waves of planned obsolescence under the banner of environmental friendliness, thus legitimating consumerism and maintaining profitability.[2]

But at a deeper level corporations construct the problem of the environment in a way that is the polar opposite of that in which greens typically think about reform—namely, the problem of how to remake nature in ways that are consistent with sustainable profitability and capital accumulation. "Remaking nature" means more access to nature as "tap" and "sink," which has political and ideological as well as economic and ecological dimensions, for example, the assault on the lives of indigenous peoples. Remaking nature also means reworking or reinventing nature (the political and ideological aspects of which are also important), for example, by means of "even-age industrial plantations" of pine and fir in the U.S. Southeast and

Northwest—a monoculture that has been called "forestry's equivalent to the urban tower block"; genetic alteration of food to reduce crop losses and increase land yields; microorganisms used in the semiconductor industry to "eat" toxic wastes; and genetically altered ragweed plants that clean soil contaminated by lead and other metals.[3] However, each innovation has its potential dangers: plantation forestry destroys biological diversity and genetic changes in food crops and the use of microorganisms to reduce costs contain unknown biological dangers. Here we enter a world in which capital does not merely appropriate nature, then turn it into commodities that function as elements of constant and variable capital (to use Marxist categories), but rather a world in which capital remakes nature and its products biologically and physically (and politically and ideologically) in its own image.[4] A precapitalist nature is transformed into a specifically capitalist nature. And just as the labor movement forced capital to move from a mode of absolute surplus value to one of relative surplus value production, for example, from lengthening work hours to reducing the costs of wages, so the green movement may be forcing capital to end its primitive exploitation of precapitalist nature by remaking nature in the image of capital—also to lower the costs of capital, especially the costs of reproducing labor power (or the cost of wages).

Seen this way, nature would become unrecognizable as such, or as most people experience it. It would be, rather, a physical nature treated as if it is governed by the law of value, the process of capitalist accumulation through economic crisis, like the production of pencils and fast foods. Discourse theory will then have as much to say about the problem of sustainability as do political economy and ecological science. The reason is that the capitalist project to remake nature—still in its infancy—is also a project to remake (pretendedly) science and technology in the image of capital. What this image is or will be depends on complex issues of representation, images of nature, and problems of social solidarity, legitimation, and power within scientific and university communities.

3. Demand Crisis: Expansion and Consumption

A systematic answer to the question "Is an ecologically sustainable capitalism possible?" is "Not unless and until capital changes its face in ways that would make it unrecognizable to bankers, money managers, venture capitalists, and CEOs, looking at themselves in the mirror today." This assertion, widely rejected by national politicians and spokespeople for big business, requires for its justification a brief account of how capitalism works, why it works when it works, and why it does not work when it does not.

Until the rise of ecological economics—which, despite precursors dating back more than a century, is still at the fringes of the profession—economists discussed the sustainability of capitalism in purely economic terms, for example, money capital, investment and consumption, profits and wages, costs and prices. The physical or material world appeared in models of economic growth in just two guises: first, in the form of location and rent theory; second, in the concept of the "accelerator," or the amount of physical product that new productive capacity can be expected to produce (e.g., at a given rate of utilization, so many machines are needed to produce so many refrigerators).

From an economic point of view, sustainable capitalism must of necessity be an expanding capitalism (and represented as such). A capitalist economy based on what Marx called "simple reproduction" and what many greens call "maintenance" is a flat impossibility—the (unpaid) maintenance work of domestic labor and (paid) work organized by the state excepted. There is little or no profit in maintenance; capitalistic sustainability depends on profits. A positive overall rate of profit means growth of total product ("gross domestic product," as measured in capitalist national income accounts). Profit is the means of expansion, for example, of new investments and technologies. Profit also functions as an incentive to expand. Profit and growth are thus means and ends of one another, content and context, as it were, and the average money manager does not really see or care about the difference between them. While there are many variations of economic growth theory, all presuppose that capitalism cannot stand still, that the system must expand or contract, that is, that it is both crisis-ridden and crisis-dependent, and, in the last analysis, that it must "accumulate or die," in Marx's words.[5]

In the simplest (and most simpleminded) model of capitalism, the rate of growth or rate of accumulation of capital depends on the rate of profit.[6] The higher the profit rate (everything else being the same), the more sustainable is capitalism. A negative profit rate spells economic trouble: at the least, a recession, at the most, a general crisis, deflation of capital values, and depression. According to this model, anything or anybody that interferes with profits, new investment, and expanding markets threatens the sustainability of the system, that is, an economic crisis with unknown and unknowable economic, social, and political consequences.

In traditional Marxist theory, capital is its own worst enemy. Capital threatens its own profitability because of what Marx called the "contradiction between social production and private appropriation." This means that the greater the degree of capital's political power over labor, the greater will be the degree of exploitation of labor (or the rate of surplus value), and the more potential profits will be produced. However, precisely for this reason, the greater will be the difficulty of realizing these potential profits in the market, or to sell goods at prices reflecting costs of production plus the average profit rate. Here we identify the contradiction between capital's political power and the ability of the capitalist economy to work smoothly (or, at the limit, to work at all). This "first contradiction of capitalism" (or "realization" or "demand crisis") states that when individual capitals attempt to defend or restore profits by increasing labor productivity, speeding up work, cutting wages, and using other time-honored ways of getting more production from fewer workers, meanwhile paying them less, the unintended effect is to reduce the final demand for consumer commodities. Fewer workers, technicians, and others in the labor process produce more, who by definition are able to consume less, absent a deflation of prices. Thus, the greater the produced profits, or the exploitation of labor, the smaller the realized profits, or market demand—all other things being equal. Of course, other things are never equal: government budget deficits, mortgage and consumer credit, business borrowing, and an aggressive foreign trade policy, among other things, may buoy up demand to keep capital "sustainable."

Today, a sustainable economy presupposes a global political economic system able to identify and regulate this "first" or "internal" contradiction of capitalism. This means first and foremost the capacity for macroeconomic regulation on a global scale, or, at least, between the industrial powerhouses of the Group of Seven (G7), that is, an international Keynesianism of the type installed in the leading national economies from the 1950s through the late 1970s. Defined in this immediate and practical way, world capitalism may be much less sustainable than most economists think. First, the systems of national Keynesian regulation have weakened or self-destructed since the late 1970s. Second, the central role of the United States in the global economy until the post-cold war period—as a kind of world cash register—has ended. This means that, until the weak recovery from the recession of 1990–1991, the U.S. economy was driven by consumer and military spending and private and public borrowing. The present recovery, however, is the first since 1876 to be led by export spending, with investment spending a close second. All of Germany's recent recoveries have been export-led, and the German government has said that any recovery from its present economic doldrums will be export-driven. If and when Japan recovers from its present economic troubles, it will do so because exports will expand faster than domestic consumption, investment, and government spending. Finally, all of the so-called newly industrializing economies are export-driven. In a period in which a consumerist United States can no longer absorb the world's surplus commodities, global macroeconomic management of a Keynesian type will be needed to avoid a general deflation and depression.

In fact, there is a sort of global macromanagement composed of the finance ministers of G7, the International Monetary Fund, the Bank for International Settlements, the World Bank, and the General Agreement on Tariffs and Trade. This quasi-global capitalist state, however, is in the hands of big capital in general and finance capital in particular. Hence, with the exception of G7's attempts to lower interest rates and stimulate demand in countries with export surpluses (especially Japan), the global state follows an anti-Keynesian policy, one that forces individual capitals and whole countries to cut costs, increase efficiency, and lower government spending, respectively, without a second thought about the effects of this policy on capital overproduction on a global scale—of the type Marx identified long ago—not to speak of the dangers of bitter trade wars, creative forms of beggar-my-neighbor policies, growing social decay, political instability, and regional trading blocks. Put another way, there is no global parliament to pass minimum wage laws and protective legislation, no World Ministries of Labor or Social Welfare, no World Ministry of Environment, no legitimate power spreading Keynesian economic literacy on an international scale. Instead, in the United States, for example, ex-President Bush demanded that the United States become an "export superpower" and President Clinton's economic advisers called for an "increasingly aggressive" export policy.

The prospects of global regulation today, organized in a truly cooperative spirit, are as poor as those of national regulation during the crisis of the 1890s, namely, zero. In those days nationalist policies of dumping, monopoly, and colonialism helped to create two world wars of imperialist rivalry and the Great Depression. Superficially, there might be two mitigating factors today: one is that Europe is an economic entity; France, for example, wants to join, not fight, Germany, economically. The other is that capital is no longer national, but rather global in scope, hence theoretically more open to global regulation. But G7 has done a poor job of macroeconomic regulation to date, and global finance capital and the rentier class living off interest on the huge piles of debt accumulated in the 1970s and 1980s are powerful enough to prevent governments from reflating their economies.

4. Cost Crisis: Conditions of Production

Today, this kind of economic thinking, while still valid, is one-sided and limited (as, in fact, it always was). The reason is that it presupposes limitless supplies of what Marx called "conditions of production." This traditional model presupposes that capitalism can avoid potential bottlenecks on the "supply side," that growth is demand-constrained only. However, if the costs of labor, nature, infrastructure, and space increase significantly, capital faces a possible "second contradiction," an economic crisis striking from the cost side. Famous examples include the English "cotton crisis" during the U.S. Civil War; wage advances in excess of productivity in the 1960s; and the "oil shocks" of the 1970s. However, here we are concerned with phenomena that are much more structured or generic than these isolated examples by themselves would suggest.

Cost-side crises originate in two ways. The first is when individual capitals defend or restore profits by strategies that degrade or fail to maintain over time the material conditions of their own production, for example, by neglecting work conditions (hence raising the health bill), degrading soils (hence lowering the productivity of land), or turning their backs on decaying urban infrastructures (hence increasing congestion costs).

The second is when social movements demand that capital better provides for the maintenance and restoration for these conditions of life; when they demand better health care; protest the ruination of soils in the name of "environmental protection"; and defend urban neighborhoods in ways that increase capital costs or reduce capital flexibility—to stay with the same three examples. Here we are talking about the potentially damaging economic effects to capitalist interests of labor movements, women's movements, environmental movements, and urban movements. This problem of "extra costs"—and their threat to profitability—obsesses mainstream economists and capitalist ideologists; the leaders of labor and social movements are rarely willing to discuss it in public, however.

In the real world, both types of cost-side crises combine and intermingle in complex and contradictory ways that no one has ever systematically theorized. For example, from a quantitative standpoint, no one knows how much urban congestion costs are the result of capital's celebration of the automobile and its neglect of urban mass transport, and how much the effects of community struggles to keep freeways from scarring their neighborhoods.

We need a more refined theoretical approach to the problem of what Polanyi called "land and labor." Marx inadvertently supplied a start for such an approach with his concept of "conditions of production."[7] Conditions of production are things that are not produced as commodities in accordance with the laws of the market (law of value) but which are treated as if they are commodities; that is, they are "fictitious commodities" with "fictitious prices." According to Marx, there are three conditions of production: first, human labor power, or what Marx called the "personal conditions of production"; second, environment, or what Marx called "natural or external conditions of production"; third, urban infrastructure (we can add "space"), or what Marx called "general, communal conditions of production." The fictitious price of labor power is the wage rate, and that of environmental and urban infrastructure and space is rent. Given that wage and rent theory are not and cannot be based on costs of production, it is understandable that both bourgeois and Marxist economic accounts of wages and rents are the least developed and least satisfying topics in the entire economics literature.

Sustainable capitalism would require all three conditions of production to be available at the right time and the right place and in the right quantities and the right qualities, and at the right fictitious prices. As I noted, serious bottlenecks in the supply of labor power, natural resources, and urban infrastructure and space threaten the viability of individual capital units—and even of entire sectoral or national capitalist programs. If generalized, these bottlenecks would thus threaten the sustainability of capitalism by driving up costs and impairing the flexibility of capital. "Limits to growth" thus do not appear, in the first instance, as absolute shortages of labor power, raw materials, clean water and air, urban space, and the like, but rather as high-cost labor power, resources, infrastructure, and space. This immanent threat to profitability leads the state and capital to attempt to rationalize labor markets, supplies, markets for fuel and raw materials, and urban and rural land-use patterns and land markets to reduce costs of production.[8]

Supply-side bottlenecks or shortfalls pose especially difficult problems for capitalist enterprises and policymakers when the economy is weak, or when it faces a demand-side crisis or fresh competition from other countries. Stagnant or falling profits force individual capitals to attempt to reduce the turnover time of capital, that is, to speed up production and reduce the time that it takes to sell their products. This obsession with making money faster and faster to compensate for low or falling profits runs up against, for example, union-organized labor markets, OPEC-influenced oil markets, and traditional agriculture's defense of "inefficient" uses of the soil and water. On the one hand, money capital seeks more of itself faster and faster; on the other hand, what Polanyi called "society" and what capital regards as out-of-date patterns of land and labor utilization, and land and labor markets, combined with the resistance to capitalist rationalization by labor and social movements, all constitute themselves as obstacles, or "barriers to overcome." At the very least, capital must deal with social indifference and inertia.

One of capital's solutions to this dilemma, at least in the short term, is as simple as it is economically self-destructive. Money capital abandons the "general circuit of capital"—that is, the long and tedious process of leasing factory space, buying machinery and raw materials, renting land, finding the right kind of labor power, organizing and implementing production, and marketing commodities—and finds its way into speculative ventures of all kinds. Money capital, based on the expansion of credit, or money that cannot find outlets in real goods and services, jumps over society, so to speak, and seeks to expand the easy way—in the land, in stock and bond markets, and in other financial markets. Hence the present economic anomaly: the value of claims on the surplus or profits grows at the same moment that the real value of fixed and circulating capital stagnates or declines. This tends to make a bad economic situation worse, for it causes growing indebtedness and the danger of a financial implosion. It also tends to worsen ecological and other conditions of production: as financial interests assume hegemony over productive interests, the environment tends to be neglected even more.

During earlier periods of capitalist development, and defined in functional terms, there was sufficient precapitalist labor power, untapped natural wealth, and space. This was true, in fact, and also in terms of the perceptions of early generations of bourgeoisies. The (fictitious) prices of labor power, natural resources, and space were thus held in check. Nor did environmental movements or urban movements raise political and social barriers to capital that capital itself (with the help of imperialism and state oppression) could not overcome. Over time, capital seeks to capitalize everything and everybody; that is, everything potentially enters into capitalist cost accounting. For millennia, human beings have been "humanizing" nature or creating a "second nature," and this has often been destructive, for example, deforestation and drought–flood cycles under the Roman plantation system; the devastating ecological consequences of the Punic Wars; and soil depletion and water scarcity in Mayan civilization. But in capitalist social formations, this second nature is commodified and valorized at the same time as it is being degraded. From the point of view of those who want capitalism to be ecologically sustainable, this is when problems start to appear. Labor markets become tight, and the North has to rely on imported labor from the South—with all the attendant economic and social costs and problems. Examples include the economic cost of settling newcomers who use a different language and the social costs of a resurgence of racism. Raw materials and unpolluted commons become scarce, driving up what Marx called the "costs of the elements of capital," for example, U.S. domestic oil and gas; trees and lumber; and supplies of clean water. And last but not least, urban infrastructure and space become scarce, creating rising congestion costs, higher ground rents, and greater pollution costs. Los Angeles is a good example; Mexico City and Taipei are better ones.

Thus the capitalization of the conditions of production in general and the environment and nature in particular tend to raise the cost of capital and reduce its flexibility. As I have noted, there are two general reasons. First, a systemic reason, namely, that individual capitals have little or no incentive to use production conditions in sustainable ways, especially when faced with economic bad times of capital's own making. Second, precisely for this reason, labor, environmental, and other social movements challenge capital's control over labor power, the environment, and the urban (and, increasingly, the rural as well, especially in the South). Examples in the United States include regional Toxics Coalitions, occupational health and safety and right-to-know struggles, direct action to save wild rivers and original-growth forests, and antifreeway and antidevelopment movements.

In summary, these lines of thinking suggests that there are two, not just one, contradictions of capitalism, two, not just one, types of economic crisis, and two, not just one, types of crisis resolution. The "second contradiction" of capitalism results in economic crisis that strikes not from the demand side but from the cost side. Since the 1960s many economists, leftist and right-wing alike, have defended the thesis that cost-side (as well as demand-side) crises characterize the late 20th century, for example, the cost-push crisis of profitability due to wage increases, struggles against productivity, welfare state expansion, high oil prices, overregulation of business, and so on.

Put simply, the second contradiction states that when individual capitals attempt to defend or restore profits by cutting or externalizing costs, the unintended effect is to reduce the "productivity" of the conditions of production, and hence to raise average costs. Costs may increase for the individual capitals in question, other capitals, or capital as a whole. For example, the use of pesticides in agriculture at first lowers, then ultimately increases, costs as pests become more chemi-

cal resistant and also as the chemicals poison the soil. Permanent-
yield monoforests in Sweden were expected to keep costs down, but
it turned out that the loss of biodiversity over the years has reduced
the productivity of forest ecosystems and the size of the trees. In the
United States, nuclear power promised to reduce energy costs. But
bad design, problems of finance, safety measures, and most of all
popular opposition to nuclear power had the effect of increasing costs.
As for the "communal" conditions of production, new highways
designed to lower the costs of transportation and the commute to
work tend to raise costs when they attract more traffic and create more
congestion. And in relation to "personal" production conditions, it
is clear that the U.S. education system, which is supposed to increase
potential labor productivity, produces as much stupidity as learning,
impairing labor discipline and productivity.

It is important to stress the idea that the conditions of production
are not produced in accordance with laws of the market. Nor does
the market generally regulate capital's access to these conditions when
and if they are produced. There must be some agency, therefore,
whose task it is either to produce the conditions of production or to
regulate capital's access to them. In capitalist societies, this agency
is the state. Every state activity, including every state agency and
budgetary item, is concerned with providing capital with access to
labor power, nature, or urban space and infrastructure. For example,
in the United States, there are the labor and education bureaucracies;
the Department of Agriculture; national and state park services; the
U.S. Bureau of Land Management and the U.S. Bureau of Reclama-
tion; and urban planning bodies and traffic authorities. Examples of
specific functions related to the three conditions of production are,
first, with respect to labor power, child labor laws and laws governing
hours and conditions of work and work safety; second, in relation
to environment, laws governing access to federal lands and regulating
coastal development and pollution; third, with respect to urban infrastructure and space, zoning laws, traffic planning, and land-use regulation. There is hardly any state activity or budgetary item that does
not concern itself in different ways with one or more conditions of
production. This also includes monetary and military functions,
which safeguard and facilitate "legitimate" access by capitalist mining
companies, banks, merchants, and other enterprises to needed resources and markets. Bush's war in the Persian Gulf is only the latest
and most dramatic example of the role of the military in capitalist
societies; the World Bank and the International Monetary Fund (at
the supranational level) are the most obvious examples of monetary
functions oriented to capitalist expansion.

5. Managing Cost Crisis

What is the solution to these cost-side crises, from the standpoint of individual capitals and also from that of capital as a whole?

The worst case is when individual capitals, faced with both higher costs and lower demand, cut costs even more, thus intensifying both the first and second contradictions. But this result is not the only possibility. As I noted, in relation to the environment, there are many examples of individual capitals responding to green consumerism, such as the public demand to reduce waste and recycle by finding new uses for waste products; and also many examples of companies that upgrade their capital equipment when forced to reduce their pollutants and other companies that specialize in environmental cleanup.

The best solution for capital as a whole (though not for society nor even for "nature"; this would presuppose a logic of reciprocation rather than the logic of capitalist exchange value) is to restructure the conditions of production in ways that increase their "productivity." Since the state either produces or regulates access to these conditions, restructuring processes are typically organized and/or regulated by the state, that is, politically. Examples include banning cars in urban downtowns to lower congestion and pollution costs; subsidizing integrated pest management in agriculture to lower food and raw materials costs; and shifting emphasis from curative to preventative health (e.g., the fight against AIDS in the United States) to lower health care costs. However, huge sums of monies would have to be expended—to attain a real solution—to restructure production conditions in ways that restored or increased their "productivity" and thus lower the costs of capital. Long-term productivity would be enhanced, but at the expense of short-term profits. New industries would produce environmentally friendly products, urban transport, and education systems, which (like the examples cited above) effectively lower the costs of the elements of capital and of the consumption basket, as well as of ground rent; at the same time, the level of aggregate demand would be increased, attacking the first contradiction in potentially noninflationary ways. (By contrast, if new systems of forest management, pollution control spending, urban planning, and so on, have no effect on costs, the result is an increase in effective demand and inflation or a reduction in profits.)

So much for the idea of sustaining capitalism; practice is another question. In liberal democratic states the normal political logic of pluralism and compromise prevents the development of overall environmental, urban, and social planning. The logic of the state adminis-

tration or bureaucracy is undemocratic, hence insensitive to environmental and other issues raised from below. And the logic of self-
expanding capital is antiecological, antiurban, and antisocial. All three
logics combined are contradictory in terms of developing political
solutions to the crisis of the conditions of production; hence chances
of instituting a systematic "capitalist solution" to the second contradiction are remote.

Put differently, there is no state agency or corporatist-type planning mechanism in any developed capitalist country that engages in
overall ecological, urban, and social planning. The idea of an ecological capitalism, or a sustainable capitalism, has not even been coherently
theorized, not to speak of becoming embodied in an institutional
infrastructure. Where is the state that has a rational environmental
plan? Intraurban and interurban planning? Health and education planning organically linked to environmental and urban planning? Nowhere. Instead, there are piecemeal approaches, fragments of regional
planning at best, and irrational political spoils allotment systems at
worst.

Every day, therefore, new headlines announce another health
care crisis, another environmental crisis, and another urban crisis. In
many regions the ultra-image we have is of an increasingly illiterate
labor force, many of whom are homeless because of low wages and
high rents, living in fear in a polluted city, immobilized by gridlock,
and unable even to obtain clean water. This picture may not fit Rome
or New York yet, but it comes close to describing Mexico City and
New Delhi, which by any measure are parts of the capitalist world.

6. Ecological Consequences of a General Economic Depression

However sustainability is defined from an ecological standpoint, one
thing seems fairly certain. If capitalism is not sustainable in terms of
international macroeconomic regulation, there will be a global crisis,
a general deflation of capital values, and a depression. In this event,
no one knows or can know how individual capitals, governments,
and international agencies will respond.

It may be that great economic pressures from the demand side
(or the cost side or both at the same time) arising from capital overproduction (or underproduction or both) would force individual capitals
to try to restore profits by externalizing more costs, that is, by shifting
more cost to the environment, land, and communities, with states
and international agencies looking on helplessly. In fact, there is plenty

of evidence that the slow economic growth since the mid-1970s has resulted in such cost shifting, especially by transnational corporations. There is also evidence that in many cases this plan has backfired in the sense that cost shifting by one capital has increased costs for other capitals. Also, it can be shown that in many cases environmental struggles and environmental regulation have forced individual capitals to internalize costs that would otherwise fall on the environment. There is a kind of war going on between capital and the environmental movements, a war in which these movements might have the effect (intentional or not) of saving capital from itself in the long run by forcing it to deal with the negative short-term effects of cost shifting.

There is also the possibility, however slight, that a real economic depression might be the occasion for a general program of environmental restoration. In the United States in the 1930s the New Deal created the political conditions for two types of environmental changes. The first consisted of massive efforts to restore the degraded soils of the Great Plains and the ecologically damaged south and western rangelands. In this sense, the depression was an "environmentally sound" event. The second type of environmental change consisted of even greater efforts to start up or to speed up giant infrastructure projects, such as huge dams, great bridges, and tunnels, which were indispensable for urbanization in the West and for post-World War II suburbanization everywhere in the country. Without these projects, consumerism and the culture of the automobile could not have flourished in the 1950s and 1960s. These projects in important ways helped to create the present structure of individualist consumption, which is ecologically unsound.

The next depression may make environmental conditions much worse, or it may be the occasion for vast changes in the structure of individual and social consumption, for example, green cities, integration between cities and surrounding agricultural lands, public transport that people look forward to using, and so on. Or both, to varying degrees, in different places. What will actually happen, of course, will be decided by the course of political struggle, institutional adaptation, and even types of technological innovation.

All of which is to say that environmental destruction, environmental and related social movements, government policies and budgets, policies of international bodies, and economic conditions are as interrelated as any complex ecosystem modeled by professional ecologists. Anyone trying to think about these interrelations will run up against the same epistemological and methodological problems that ecologists face when they try to model the fate of some particular species, that is, the problem of atomism and reductionism versus holism. Worse, while bald eagles and microorganisms do not organize themselves politically as social agents, people sometimes do. Hence a strict systems theory approach to the question of the ecological effects of a general depression is of questionable usefulness. In the last analysis, everything depends on the balance of political forces, and the visions of those who want to transform our relation with nature, hence our material relationships with one another—in short, with the political objectives of the environmental, labor, women's, and other social movements. "Is sustainable capitalism possible?" is in the last, as well as in the first, instance, a political question.

7. Conditions in the South

The crisis of the conditions of production is especially severe in the South, which explains the appearance of the discourse on "sustainable development" there, which has become an ideological and political battleground of growing importance. As I have noted, practically everyone uses the expression, with different intentions and meanings. Sustainability is defined by environmentalists and ecological economists to mean the use of only renewable resources and also low or nonaccumulating levels of pollution. The South may be, in fact, closer to "sustainability" seen in this way than the North is, but the North has more capital and technological resources than the South as the means to attain sustainability. Capital, of course, uses the term to mean sustainable profits, which presuppose long-run planning of the exploitation and use of renewable and nonrenewable resources, and of the "Global Commons." Ecologists define sustainability in terms of the maintenance of natural systems, wetlands, wilderness protection, air quality, and so on. But these definitions may have little or nothing to do with sustaining profitability. In fact, there is generally an inverse relation between ecological sustainability and short-term profit. The "sustainability" of rural and urban existence, the worlds of indigenous peoples, the conditions of life for women, and safe workplaces are also inversely correlated with the sustainability of profits—if the history of the long economic crisis of the late 20th century is any guide.

Independent of the question of the desirability of the South following the industrial, consumerist path of the North is the possibility of it doing this. Industrial capitalism in India, Brazil, and Mexico (to take three examples) occurs at the expense of vast poverty and misery and also erosion of ecological stability, however this expression is defined. East Asia is doing well economically, and some Southeast

Asian countries are doing even better (in terms of growth of gross domestic product), but these regions have yet to prove that they can be industrial powerhouses and also pay good wages and provide decent working conditions, progressive social policies, good conditions of urban life, and meaningful environmental protection. Most of the rest of the South (including the North's and East Asia's internal colonies) is an economic, social, and ecological disaster zone. There are many barriers to capitalist development in the South, for example, weak markets, due to hugely unequal distribution of wealth and income; the absence of agrarian reform favoring small and middle farmers; and instabilities in the demand for, and supply of, raw materials. Also, there are problems relating to foreign debt and balance of payments crises, not to speak of maintaining ruling blocks of propertied interests and stable governments. These problems exist independently of the state of ecological conditions in particular and the conditions of production in general. Needless to say, this situation creates permanent social and political instability; new migration patterns to the North; more economic and ecological refugees; and so on—which, in turn, spell continued trouble in the North.

8. Political Possibilities

The majority of the center–right and rightist governments that have been governing most of the world since the late 1970s and early 1980s are incapable of steering capitalist development in ways that improve the conditions of life, labor, the cities, or the environment. These governments are too intent on expanding the "free market" and the international division of labor; deregulating and privatizing industry; forcing economic "adjustments" on the South and "shock therapy" on the old socialist countries—hence marginalizing up to half the population of some third world countries, and pretending that the "market" and neoliberalism generally will solve the growing economic crisis. But things are likely to get much worse before they get better, especially in the South.

Meanwhile, there has been a growth of various green and "red green" movements in different countries. Politically speaking, arguably the most developed is New Zealand's Alliance.[9] A few labor unions in some countries are addressing environmental issues more seriously. Conversely, environmental movements are addressing economic and social issues that 5 or 10 years ago they ignored or downplayed. In multiple forms, labor and feminist movements, urban movements, environmental movements, and movements of op-

pressed minorities have organized themselves around the general issues of the conditions of life. While the prospects for a sustainable
capitalism are dim, there may be hope for some kind of ecological
socialism—a society that pays close attention to ecology along with
the needs of human beings in their daily life, as well as to feminist
issues, antiracism, and issues of social justice and equality generally.
Globally, it is around these issues that there is movement and organization, agitation and action, which can be explained in terms of the
contradictions of capitalism and the nature of the capitalist state I
discussed above.

Politically, this means that sooner or later labor, feminist, urban,
environmental, and other social movements need to combine into a
single powerful, democratic force—one that is both politically viable
and also capable of radically reforming the economy, polity, and
society.[10] Individually, social movements are relatively powerless in
the face of the totalizing force of global capital. This suggests the
need for three general and related strategies.

The first is the self-conscious development of a common or
public sphere, a political space, a kind of dual power, in which minority, labor, women, urban, and environmental organizations can work
economically and politically. Here there could be developed not the
temporary tactical alliances among movements and movement leaders
that we have today, but strategic alliances, including electoral alliances. A strong civil society, defining itself in terms of its "commons," its solidarity, and its struggles with capital and the state, as
well as of its democratic impulses and forms of organization within
alliances and coalitions of movement organizations—and within each
organization itself—is the first prerequisite of sustainable society and
nature. The second is the self-conscious development of economic
and ecological alternatives within this public sphere or "new commons"—alternatives such as green cities, pollution-free production,
biologically diversified forms of silviculture and agriculture, and so
on, the technical aspects of which are increasingly well known today.
The third is to organize struggles to democratize the workplace and
the state administration so that substantive contents of an ecological,
progressive type can be put into the shell of liberal democracy. This
presupposes that the movements not only use political means to economic, social, and ecological goals, but also agree on political goals
themselves, especially the democratization of some national and international state apparatuses, and the elimination of others.

These ideas may seem to be as unrealistic as that of an ecological
capitalism. Perhaps this is the case. But we need to remember that
while the existing structures of capital and the state do not seem to be

capable of anything more than occasional reforms, social movements worldwide grow every day—hence it is possible that at some point there will be a general social and political crisis, as the demands of these movements clash with existing profit-oriented economic and political structures. If this point is reached, all kinds of "social morbid forms" will appear. Some will say that this is precisely what is happening today—that the social and political fabrics are unraveling, and that the resurgence of racism, nativism, discrimination against foreign workers, male and antienvironmental backlash, and other reactionary trends and tendencies are becoming increasingly great dangers. Others link the revival of right-wing populism and reaction to rightward shifts in the political and economic mainstream. There are other analyses of the current world political situation—including the line that the globe is witnessing a rebellion of the well-to-do against the demands of the poor, the welfare state, redistributive economic policies, and the like—a war of the rich against the poor. Or all of the above may be true. Whatever the case(s), from the standpoints of progressives, red or left greens, and feminists, the last thing in the world we need is factionalism, sectarianism, "correct lineism"—instead, we need to scrutinize critically all time-worn political formulae and develop an ecumenical spirit, and to celebrate our commonalities or "new commons" as well as our differences.

Notes

1. *New York Times*, August 12, 1992.
2. See, for example, Simon Fairlie, "Long Distance, Short Life: Why Big Business Favours Recycling," *Ecologist*, 22, no. 6, November–December 1992, pp. 276–283.
3. On forestry, see Edward Goldsmith et al., *The Imperialist Planet* (Cambridge, Mass.: MIT Press, 1991), p. 94. Most timber in the United States is produced on industrial plantations. Examples of genetic modifications now abound, and there exist modified strains of almost all staples either in laboratory or in commercial use—corn, rice, soybeans, and many other foods, including a potato that kills one of its own pests, the Colorado potato beetle, by emitting a protein fatal to the insect. Wheat has been experimentally genetically altered by the University of Florida and Monsanto Co. to increase yields. They introduced a foreign gene into wheat that produces an enzyme that makes many herbicides harmless to the wheat. Of course, the gene introduced into the wheat is a trade secret (*New York Times*, May 28, 1992).
4. No more is it only a question of capital appropriating what is already found in nature, then breaking it down and recombining its elements into a commodity, but rather of creating something that did not exist before.

There is no hard line between the two; nevertheless, there is a qualitative difference once you compare the extremes.

5. All growth theories presuppose certain relationships between the "real" and money economies, physical production and incomes, and increases in investment and consumption goods, on the one hand, and profits and wages, on the other hand. Before the development of ecological economics, the question "What exactly is growing?" was relatively neglected. Today, more economists are willing to admit that growth not only includes some vector of outputs (commodities, services, increments of durable stocks of goods), but also outputs of "wastes" and increments of stocks of durable wastes. This complicates an already complex and arbitrary system of income accounting.

In classical terms, disproportionalities between the investment/consumption good and profit/wages ratios can cause economic trouble ("disproportionality crises"). The main type of crisis inherent in capitalism, however, is a "realization crisis." Marxists regard capitalism as "crisis-ridden." But the system is also "crisis-dependent" in the sense that economic crises force cost cutting, "restructuring," layoffs, and other changes that make the system more "efficient," that is, more profitable. Marx wrote that "capital accumulates through crisis," meaning that crises are occasions for the liquidation of some capitals, and also the appearance of new capitals and reorganization of old capitals, not to speak of the diffusion of new and more "efficient" technology throughout the system, such as computerization (see below).

6. "Most simpleminded" in part because while there is a general tendency for the rate of profit in different industries to become roughly comparable (via movement of capital away from low-profit and toward high-profit sectors), profit rates vary widely from industry to industry, even from capital unit to capital unit. There are many reasons for this, among which (and arguably the most important) is that big capitals not only appropriate larger profits defined in absolute or total terms than small capitals, but also the former "earn" a higher profit rate than the latter. The reason is that small capitals typically cannot compete with big capitals, while big capitals can compete with small capitals (as well as with each other).

7. "Inadvertently" because Marx used the concept of "conditions of production" in different and inconsistent ways; he never dreamed that the concept would or could be used in the way that I will use it in this chapter; and no one could have used the concept in this way until the appearance of Karl Polanyi's *The Great Transformation* (New York: Farrar and Rinehart, 1944). The "land and labor" theme has, of course, been taken up in other ways by critical writers concerned with peasant subsistence, indigenous or "First Peoples," and the nexus of women, capital, and the environment—for example, Claudia von Werlhof, "On the Concept of Nature and Society in Capitalism," in Maria Mies, ed., *Women: The Last Colony* (London: Zed Books, 1988), pp. 92–112.

8. This rationalization also includes reprivatization defined as a shift from paid labor to unpaid labor in the home and community, or the revival of self-help ideologies that throw more of the burden of reproducing labor

power and environmental and urban conditions of life onto noncapitalized domains. What Martin O'Connor calls "autonomous subsistence"—the somewhat autonomous (but simultaneously exploited) domains of household, peasant, and communal life and (re)production activities—furnishes a necessary undergirding of capitalist accumulation, and this assumes particular importance during crisis periods—on the one hand, providing degrees of freedom in capital's restructuring process, and, on the other hand, providing a subsistence possibility for people jettisoned through redundancies or social welfare cutbacks from the "formal" economy. Exploration of this issue raises the larger subject of how to theorize this articulation between capitalist and subsistence domains—whether and how, for instance, unpaid domestic labor constitutes the exploitation of women by men, functions as a subsidy to capital, and so on—questions that have been much debated by feminists, Marxists, Marxist feminists, and ecofeminists since the 1970s.

9. See Wayne Hope and Joce Jesson, "Contesting New Terrain: Red Green Politics in New Zealand," *CNS*, 4(2), no. 14, June 1993, pp. 1–17. The strongly developed and high-profile green political movement in (former West) Germany has had only limited success in building alliances with the labor left; see John Ely, "Red Green Ecological Reconstruction in Germany: A Project on Hold," *CNS*, 2(3), no. 8, October 1991, pp. 111–126.

10. No one knows or can know when a "single, powerful democratic force" will develop, or even if it will develop at all. Very difficult questions must be answered, practically and theoretically—for example, whether the very concept of such a "force" is fatally grounded in the modernist–humanist tradition of Western political philosophy, a "liberal" tradition that has been less than truly tolerant of "difference," yet for all that remains firmly grounded in affirming individual rights vis-à-vis the state. Some believe, in Martin O'Connor's words (pers. comm.), that it is important "at this moment in time, that is, the late 20th century, to explore what it means to have a coexistence of many, somewhat discordant voices, having in common their repudiation of capital's domination, yet ill-reconciled in many other ways. This is an aspect of realism, of things being 'likely to get worse before they get better.'" The question being raised here is partly one of what constitutes a "unified" force, and partly one of whether and in what sense, if desirable, it is felt to be attainable. On the one hand, even if one affirms the possibility of concordant voices, there may not be time to work through all of the tensions and to fully and mutually hear the plurality of voices, different grounds of knowledge, and so on, existing between and within social movements as they stand today. On the other hand, the need for unity against capital and for a nonexploitative, socially just, ecological society may be too great, given the configurations of political forces today, to delay efforts at development of a unified political strategy capable of confronting global capital and the developing global quasi-state (of International Monetary Fund, World Bank, and so on).

Ecology and Discursive Democracy: Beyond Liberal Capitalism and the Administrative State

John S. Dryzek

1. Introduction

Today, any credible political–economic vision must address the challenge presented by ecological problems. "The environment" can no longer be thought of as just one issue among many. Ecological problems are sufficiently widespread and serious to constitute an acid test for all actual and proposed political and economic arrangements, and for all processes of institutional reconstruction, be they incremental or revolutionary. In this chapter I will argue that currently dominant institutions fail this test, and, further, that the ensuing contradiction and confusion create a space for discursive and democratic alternatives, which in turn promise enhanced possibilities for effectively meeting the ecological challenge.

Rather than giving an exhaustive taxonomy and evaluation of existing institutional mechanisms and processes,[1] here I will focus on currently dominant arrangements in the Western world and on what might replace them. These arrangements can be characterized in terms

Professor of Political Science, University of Oregon. An earlier version of this chapter was prepared for the Workshop on Ecology, Committee on the Political Economy of the Good Society, American Political Science Association Meetings, San Francisco, 1990. I am grateful for extensive comments and criticisms by the other participants in the workshop—James O'Connor, Robert Paehlke, Kenneth Peter, and Langdon Winner—and also to Hal Aronson for his editorial help. First published in *CNS*, *3(2)*, no. 10, June 1992, pp. 18–42.

of a nexus of capitalism, liberal democracy, and the administrative state. The initial question is: To what extent can these institutions—in isolation or in combination—cope with the ecological challenge?

I will argue that the three institutions are each thoroughly inept when it comes to ecology, that any combination of them can only compound error, and also that any redeeming features are to be found only in the possibilities that they open up for their own transformation. Such a summary judgment renders the study of possibilities beyond the institutional status quo interesting. But rather than detailing and evaluating all such possibilities, I will engage in a more parsimonious style of analysis. Specifically, I will attempt to determine the kinds of political and economic structures that the collision of the institutional status quo with environmental imperatives allows, and makes attractive. Additionally, I will consider the kind of democracy (if any) the age of ecology permits, or requires.

2. Capitalism

Capitalism today is showing an environmentally friendly face. Weyerhauser is "the tree-growing company" not the "old-growth clearcutting company," which it used to be. Proctor and Gamble encloses an environmental leaflet (printed on nonrecyclable paper) in packages of its disposable diapers. Corporate publicists point to the benefits that will accrue to horribly polluted Poland and East Germany once Western corporations install modern manufacturing plants. Intellectual support comes from those who argue that the market is the best mechanism for satisfying the wants of individuals—including their wants for environmental goods. Neoconservative regulatory reformers, new resource economists, and cornucopian free marketeers all tout the virtues of privatization and decentralization in the environmental political economy.[2]

The ecological rationality of capitalism (whether monopolistic or market) is, in fact, more doubtful. The ecological case against capitalism can be summarized as follows:

First, capitalism requires economic growth, the absence of which causes reduced investment, leading in turn to general economic decline. Without growth, capitalism must confront distributional inequality, unemployment, and political instability. Not surprisingly, governments in capitalist market systems see their first concern as the promotion of growth.

If there are ecological limits to growth, then growth has to cease at some point. Defenders of the market dispute this point on the grounds that our world is infinite; the market price system will ensure that substitutes are found for any scarce resource.[3] This happy result may obtain for particular resources—for example, when whale oil ran out in the 19th century, substitutes were easily located. But there is no substitute for the assimilative capacity of the biosphere. Even if a source of unlimited chap energy (cold fusion?) could be harnessed to convert degraded nature into useful nature (e.g., by concentrating dispersed resources), no escape from this constraint is possible. For the energy so used has to be dissipated—massively—somewhere in the biosphere, threatening to warm our planet beyond tolerable bounds.[4] There is a finite supply of low entropy or order on this planet, which capitalism can only degrade. Hard evidence regarding this problem is elusive. We have no indicators of the overall state or entropic level of the biosphere. Evidence as to the condition of particular resources or environmental "sinks" can only be misleading, given the possibility of substitution.

The only escape would be if capitalism could shift to growth in economic activities that do not involve consumption of materials or environmental services. To the extent that the system becomes information capitalism—centered around the production, exchange, and dissemination of information rather than material goods—its requirement for growth might seem less obviously ecologically deleterious. However, societies moving into the information age have not erased material growth. Their levels of material consumption continue to grow, even if production is increasingly off-loaded onto other societies. Local environmental improvement resulting from this shift (such as cleaner air in Pittsburgh) should not be mistaken for general success.

Second, capitalism neglects the future. Without interest rates, capital would be free, and capitalism inconceivable. The existence of positive rates of interest means that market actors must discount anticipated future costs and benefits at the prevailing interest rate. The higher that rate, the more shortsighted the system becomes. Long-range results receive little weight.

Third, as decentralized systems governed by a logic of self-interest, markets have no mechanisms for dealing with the common property and the public goods problems they generate. Actors oriented to private profit are unconcerned about damage to third parties not directly involved in transactions, and still less so to unpriced environmental resources. The tragedy of "the commons" is that instrumentally rational actors motivated by private material self-interest in an unregulated social environment will eventually ruin resources held in common. This behavior is not only functional for the operation of capitalist market systems but actually necessary for actors wishing to prosper in these systems.

Solutions to public goods and common property problems that retain the basic structure of capitalism occasionally are sought and implemented. Examples include rearrangement of property rights to give each actor a private stake in a former commons, regulation of access to resources, and the control of harmful practices such as pollution. But all such regulation requires governmental action. Recent attempts to deal with ozone depletion, global warming, and other problems have proceeded within the political arena. Capitalism displaces onto government the environmental problems that it generates but cannot solve. However, governments operating within mixed capitalist political–economic systems may not be able to respond effectively, for reasons I will now discuss.

3. Liberal Democracy

Liberal democracy can be defined in terms of competition for elective office, the opportunity to exert popular pressure upon government through free political association, a range of individual rights against government secured through constitutional limitations, and a politics of the strategic pursuit of interests defined in the private realm. During the last 20 years, more progress has been made in dealing with environmental problems in the world's liberal democracies than in countries with all other kinds of political systems. This is done mainly through the familiar mechanisms of interest group politics. Environmentalists have joined the play of interest groups, and a spate of environmental legislation has been the result. There is little doubt that liberal democracy resolves ecological problems more effectively than the market with which it coexists, and which passes on to liberal democracy the environmental problems it generates but cannot resolve. Nevertheless, the spectacular failure of liberal democracy to cope with energy problems in the 1970s, issues related to nuclear wastes, and the continuing impasse on old-growth forests in the Western United States suggests that there are limits to the problem-solving capacity of liberal democracy. These limits are fourfold.

First, the distribution of power in liberal democratic systems is inevitably skewed. Business always has a "privileged" position due to the financial resources available to it, government officials' need for business cooperation in implementing policies,[5] and government's fear of an investment strike and economic downturn if it pursues antibusiness policies.[6] And business, of course, will normally push for policies that favor its own profits, rather than ecological values. Waves of public concern for environmental values, such as the one in the United States that peaked around Earth Day in 1990, typically break when they threaten the privileged position of business. Election day in November 1990 saw environmental initiatives failing wholesale as business mobilized a public relations blitz declaring that these initiatives would take away people's jobs.

Second, liberal democracies identify and disaggregate environmental problems based on the particular interests of affected parties. These particular interests do not necessarily add up to any general ecological interest, even if some of them are environmentally inspired. American liberal democracy's spectacular confusion during the 1970s energy crises shows how particular interests can run wild and prevent coherent action.

Third, the political currency of liberal democracy consists of tangible rewards to identifiable interests. While this currency may be functional in terms of the distribution of the costs and benefits of government activity, it is of no use when it comes to large-scale, nonreducible ecological problems.

Fourth, the time horizon in liberal democracy is often no longer than that in the market. Short-term problems receive the most attention, and the next election often acts as the outer limit of foresight. Projects with anticipated long-range benefits are sometimes funded—fusion research is an example—but the funding comes not as a result of foresight, but to placate the particular constituencies who benefit from this funding in the here and now.

Fifth, liberal democracy, no less than the market, is addicted to economic growth. If growth ceases, then distributional inequities become more apparent, and the "political solvent"[7] of economic growth is no longer available. This fear of economic downturn means that liberal democracies are imprisoned by the market's growth imperative.[8]

4. The Administrative State

Mainstream politics today consists of an uneasy juxtaposition of liberal democracy and the administrative state—uneasy because administration proceeds by an instrumentally rationalistic logic, while liberal democracy's logic is political and interactive. Inasmuch as its contribution can be isolated from that of liberal democracy, administration's distinctive claim to ecological rationality rests on its purported embodiment of common purpose, neutral expertise, the capability to make sense of complex problems, and the will and authority to effect solutions to these problems—if necessary against the will of recalcitrant actors and interests such as polluters and despoilers.[9] Strengthened central administration is attractive to a number of environmental analysts, including ecological Hobbesians,[10] structural reformers with coordinated environmental management in mind,[11] socialist planners,[12] and Burkean conservatives who would like to see mass restraint in environmental demands inculcated by some wise elite.[13]

However, administrative systems are themselves imprisoned in three ways:

First, these systems are highly constrained in their responses to problems. Compliance on the part of subordinates to goals established at the apex of the hierarchy is far from automatic, undermining any claim to common purpose. Further, bureaucratic organizations can only be programmed to perform a limited range of routines, which are unlikely to be adequate for complex and variable environmental problems. Just as the military is always prepared to fight the last war, so adding environmental instructions to the mandates of agencies is not going to change their ways. Five-year plans in Eastern Europe regularly incorporated environmental targets, but to no practical effect. Large, centralized organizations produce large, centralized solutions to problems. Thus, the "water buffalo" agencies of federal, state, and local government in the U.S. West respond to drought with ever more extravagant schemes for capturing distant water, rather than considering more appropriate distributions of people and economic activity. In addition, hierarchical organization means that failure or misinterpretation at any point in the chain of command can be devastating. A burgeoning political science literature on policy implementation demonstrates that successful implementation is the exception rather than the rule, in environmental matters no less than elsewhere.

Second, administrative rationality cannot cope with truly complex problems. Problem disaggregation through administrative division of labor means that aspects of complex problems are artificially separated from one another. As a result, actions that look like solutions from the perspective of one subunit in reality involve displacing a problem elsewhere. This displacement can occur across space—for example, when a toxic waste dump is cleaned up through the transfer of its contents to another dump. More insidiously, problems can be displaced across the media—for example, when the Environmental Protection Agency "solves" an air pollution problem by mandating the capture of the pollutant in a toxic sludge. Displacement across time is also a possibility, as when long-lived nuclear wastes are placed in a hole in the ground.

Administrative problem disaggregation also undermines any claims to *neutral* expertise, for different administrative units typically contain different kinds of specialists committed to different values. For example, administration of old-growth forests in the Pacific Northwest pits two bureaus of the Department of the Interior against one another. The Fish and Wildlife Service is sensitive to ecological diversity and was responsible for placing the northern spotted owl and the marbled murrelet on the endangered species list. The Bureau of Land Management is more attuned to the commercial yield of logs from the land it controls.

Third, hierarchical systems necessarily obstruct the free transmission of information that is essential to the effective solution of nonroutine problems. In the context of ecological problems no less than elsewhere, verified truth relevant to policy does not exist. Theories in the hands of the apex of administrative structure must always err to a greater or a lesser degree. But hierarchy resists the institutionalization of trial and error that this recognition necessitates. The archetype of an effective problem-solving community is decentralized, such that good argument rather than hierarchical authority prevails.[14] Hierarchy may be adequate for the coordination of routine tasks, but not for complex and variable problems.

All that remains of administration's claim to ecological rationality is its undeniable will and authority to impose solutions. My arguments suggest that the solutions so imposed will rarely contribute to the amelioration of environmental crisis.

5. Compounding Error

The three mechanisms that I have discussed do not exist in isolation from one another, and therefore should not be evaluated without taking into account the ramifications of their interactions. One can imagine productive combinations, for example, Paehlke's argument that the ability of the administrative state to take the long view might mesh usefully with liberal democracy's relative openness to popular pressure, and so produce results impossible for either structure in isolation, presumably by enshrining popular consensus on environmental values into commitments extending beyond the time horizons of elected officials.[15] But this combination would still have to overcome all the other shortcomings of liberal democracy and administration—shortcomings rooted in the logic of these institutions.

Though productive combinations of existing institutions are conceivable, I would argue from an ecological vantage point that these combinations generally compound error, rather than correct it. So the capitalist market "imprisons" both liberal democracy and the administrative state by ruling out any significant actions that would hinder business profitability. Liberal democracy and administration both have a problem-solving logic that proceeds by analytic decomposition, the former according to the concern and weight of particular interests, the latter by the analytical administrative mind. Their combination preserves the basic idea of decomposition, but in combining two different logics the result is likely to be less rational than either in isolation, for each logic will triumph at particular unpredictable points. Above all, these extant systems are all manifestations of a particular kind of post-Enlightenment rationality. All embody an instrumental-analytic orientation to problem solving, hence all have the tendency to mistake problem displacement for problem resolution.

This rejection of the possibility of felicitous combinations of the three extant political–economic mechanisms short-circuits many of the contemporary debates about the environment and political economy. If we think of the three mechanisms as constituting a triangle, then many of these debates concern movements in both directions along all three sides of the triangle.

Thus Paehlke favors reducing the domain of the administrative state and increasing that of liberal democracy.[16] He notes that contrary to the projections and recommendations of ecological centralizers such as Heilbroner and Ophuls, the addition of environmental issues to the political agenda in the United States and Canada in the last 20 years has been accompanied by more openness in policy debates rather than less. This openness has taken the form of public hearings, interest group activity, right-to-know laws, public inquiries, and so forth. Heilbroner and Ophuls, of course, want to move in exactly the opposite direction on this side of the triangle. Contemptuous of the messiness, drift, and fragmentation of liberal democracy, they have no hesitation in proposing leviathan. They would expect Paehlke's proposals to exacerbate the political tragedy of the commons by increasing the access of interests claiming a share of environmental resources. Paehlke, for his part, believes that the energy to confront environmental crises is best generated by the mobilization of democratic participation, not centralized administration, and that the positive moments in the record of environmental improvement through governmental action bear him out.

If we turn to the trade-off between capitalism and the administrative state, again we find advocates of movement in both directions. Regulatory reformers want to reduce the discretionary power of administrative agencies, and to increase the role of market strategies such as systems of standards, charges, and markets in pollution rights. Some of these reformers would be happy with thorough privatization of decision-making authority, and so entrust our ecological future to capitalism. There is, of course, a long tradition of argument for movement in exactly the reverse direction. Ecological Hobbesians such as Heilbroner and Hardin are no more enamoured of the decentralized market than they are of decentralized liberal democracy. The conservation movement in the United States at the turn of the century advocated scientific expertise embodied in administration as a corrective to the irrationalities of capitalism in determining the pattern of resource use. But neutral expertise embedded in administration is a myth under complex ecological conditions, and this is especially true when one adds the moral and political dimensions of issues.

On the third side of the triangle, the most prominent arguments are those of mainstream environmental interest groups, which reject capitalism (though only in its particular manifestations, not as an overall system) in favor of liberal democratic venues for the resolution of environmental problems. On the opposite side of this argument stand public choice theorists and microeconomists who combine an exposé of the economic irrationality of interest group politics (especially with regard to the dominance of partial interests over the general social welfare) with enthusiasm for the welfare-maximizing effects of market allocation, especially once rights to private property have been well defined. One prominent school of thought here is the "new resource economics," which advocates privatization of the public lands (including national parks) in the American West as an alternative to their abuse at the hands of special interests.[17] These interests, be they timber beasts or industrial tourists, get a place at the trough and feed upon both the public treasury and the ecological endowment, all at the expense of the public interest, as a result of their success in the political game. My earlier arguments suggest that when it comes to the third side of the triangle both schools of thought are correct in their critique of the other, but neither proposes a meaningful alternative.

Thus do the debates about changing the mix of present-day mechanisms move in both directions on each side of a triangle. But if all three corners of the triangle are undesirable locations, then so probably are all coordinates within it. The question then becomes how structural possibilities beyond the triangle might be located.

6. Contradiction and Opposition

Would-be shatterers of capitalism, liberal democracy, and the administrative state have occasionally had revolution in mind—although socialist revolution would, of course, have to avoid the gross ecological failings of existing and past "Marxist" political systems. Historically, the outcomes of revolutions have generally borne little relation to the intentions of revolutionaries, and there are good structural arguments to explain why this is so.[18] The most common outcome of successful revolutions is a more potent administrative state, the kind of result that would only please ecological administrative centralizers.

Rather than speculate about grandiose possibilities for sweeping structural transformation, it seems more sensible to locate the real possibilities for change at vulnerable locations in the political economy. Such possibilities exist either where there is significant opposition to dominant structures and their imperatives, or where contradiction and confusion in dominant structures renders them vulnerable to action on behalf of some alternative institutional order. Situations where both these circumstances obtain would be especially promising. In fact, where one exists, it is likely that the other will too. As James O'Connor points out, contemporary political–economic crises can no longer be understood (if they ever could be) in terms of the working through of impersonal structural forces (such as those described by orthodox Marxists).[19] Instead, such crises are partially made up of the "reconstituted human interventions" made possible by economic and political contradictions.

Laclau and Mouffe propose a vaguely postmodern and explicitly post-Marxist perspective on the contemporary sources and significance of political resistance.[20] Dismissing the idea that social class can constitute the central basis for resistance, they recognize that capitalism and the Keynesian welfare state have spawned forms of oppressions related to gender, expertise, ecology, age, even life itself (in the "exterminism" of the military–industrial complex). Each form of oppression helps constitute subjects in different ways, so there is no single privileged oppressed subject to replace the proletariat of Marxist theory. And if subjectivity is plural, then so must be radical politics. The prescription is for a "radical and plural democracy"[21] containing a variety of self-defining struggles, though Laclau and Mouffe also seek the "articulation" or connection of different struggles.

With this stress on articulation, Laclau and Mouffe avoid the postmodern tendency to match endless variety in discourses and forms of oppression with skepticism about positive political change. Postmodernism often abjures critique altogether on the grounds that any purported solid ground for critique is in reality just another oppressive discourse. However, one can accept a plurality of struggles without subscribing to postmodern relativism and passivity. Clearly, such a plurality is not the stuff of a social revolution, or even of a tactically united social movement, but it can be the stuff of an autonomous public sphere or spheres. The public sphere can be defined as the space in which individuals enter into discourse that involves mutual respect, openness, scrutiny of their relationship with one another, the creation of truly public opinion, and, crucially, confrontation with state power.

There is, of course, no guarantee that social movements will indeed embody these virtues. Nevertheless, green, feminist, direct action, and peace groups (among others) often strive for discursive and consensual decision making, in conscious contrast to the hierarchical political style of the institutions they oppose.

Acceptance of the idea of the public sphere as an aspiration and a standard implies rejection of postmodern relativism. For this idea implies that there are criteria—those of open communication, or what Habermas called communicative rationality—that rightfully transcend the boundaries of struggles and discourses. However, any such criteria must respect the plurality highlighted by Laclau and Mouffe. Habermas is accused by his postmodern critics (e.g., Lyotard) of not respecting such diversity and implicitly favoring a homogeneous situation in which all individuals would reach the same conclusion. But communicative rationality involves respect for the reflectively held positions of others, and is defensible to the extent that it is taken as a procedural standard for political interaction that does not dictate any substantive way of life.[22]

Historically, the most well-understood public sphere is the early bourgeois one as celebrated by Habermas,[23] constituted by books, newspapers, political associations, and informal discussion in urban gathering places. Here, I want to suggest that the idea of the public sphere is central to the reconstruction of the political economy on ecologically rational lines, and also that there are good reasons why intimations of this process are now discernible.

Public spheres are emerging and expanding in the context of the kinds of struggles noted by Laclau and Mouffe. Further impetus is provided by the extent of contradiction and confusion in liberal democracy and the administrative state.[24] I argued earlier that liberal democracy and the administrative state are inept when it comes to ecology, and this may be equally true when these institutions face other issues of concern to new social movements. This perceived incapacity is one factor that leads these movements to the public sphere as a different venue for political action. And to the extent that mainstream institutions exclude certain kinds of claimants, this alternative will be a necessity rather than a choice.

The precise nature of the contemporary crises of liberal democracy and the administrative state is a matter of some dispute, as the variety in recent crisis theories attests.[25] From an ecological perspective, perhaps the most interesting aspect of the crisis in dominant institutions is their claims to rationality. In this context, O'Connor and Offe interpret the crisis of the Keynesian welfare state as a conflict between legitimation and accumulation.[26] An extensive welfare state is necessary to legitimate the capitalist political and economic order, for it soothes the discontent of those who would otherwise suffer beyond endurance from the vicissitudes of the system. Also, it generally curbs capitalism's anarchy and boom–bust potential. At the same time, the welfare state has grown to the point where it undermines the incentives necessary to make the market work—for example, by eliminating the fear of unemployment for workers and bankruptcy for capitalists. Given this conflict, there is no way for the Keynesian welfare state to perform in an administratively rational fashion, for no clear hierarchy or explicit trade-offs among goals can be established.

The consequences of this kind of crisis might involve movement toward more extensive corporatist economic management in an effort to establish a coherent compromise, at least in terms of accumulation and legitimation goals as they relate to the scope of the welfare state. But any such movement is likely only to magnify the inherent failings of the administrative state in the face of the complex problems discussed above, as well as to undermine any pretensions to democratic legitimacy. Corporatist management is by definition exclusive, and therefore in constant danger of failing to secure legitimacy. As corporatism grows, social movements are increasingly pushed away from the state and into the public sphere.

For most of their history states have had to compete successfully with other states, keep order internally, and extract the financial resources necessary to finance these activities.[27] But with time, these imperatives have undergone subtle shifts. With the advent of capitalism, the promotion of accumulation became attractive as a way to secure finance. And with the rise of democracy, internal order was translated into legitimation.

To the extent that states cannot simply export or displace ecological problems, environmental conservation will be established more firmly as an additional imperative. And this establishment can only add to the contradictions of the Keynesian welfare state. There is a clear conflict with accumulation—the deleterious environmental effects of economic growth are now well understood. And to claim that one needs the fruits of growth to pay for environmental cleanup is absurd, given that *all* of this growth would produce negative effects on the environment, whereas only a *small part* of it could be siphoned off for cleanup.

I argued earlier that neither the administrative state nor liberal democracy can cope with complex ecological problems. With this evaporation of administrative rationality, and the equivocation of politicians caught between rocks and hard spots, effective policy decisions are not readily formulated or legitimated. Students of public policy lament the seeming inability of political systems to allocate losses, even when the net benefit of a decision promises to be positive.[28] Potential losers, be they bureaucratic agencies or private interests, can mobilize to block proposals.

When those threatening to veto these proposals claim to represent citizens rather than corporations or bureaucracies, the response in many corridors of power has been to seek legitimacy in ways that involve concessions to a more participatory model of democracy (implicitly recognizing the failure of corporatism). So in the United Kingdom, the central government can no longer simply build nuclear power stations: it must hold elaborate public inquiries at which all parties are allowed (at least in principle) to have their say. Canada, the United States, and some other countries have seen enormous growth in environmental and social impact assessment, participatory models of planning, right-to-know legislation, public hearings, public inquiries, regulatory negotiation, and environmental mediation.

Such phenomena might be termed discursive designs, though I would prefer to call them *incipient* discursive designs,[29] on the grounds that they represent very imperfect approximations to ideals of free discourse (and occasionally gross violations of these ideals). Because they are associated with the state, they do not constitute *autonomous* public spheres. But, however much they may be distorted in practice, they suggest that legitimate policy decisions now require not just expertise and the backing of constitutional authority, but also informed participation by all affected parties.

7. Discursive Designs and the Public Sphere

These discursive exercises are usually sponsored by those who occupy positions of political power. Legislators, government agencies, foundations, and corporations have all sought or sponsored mediation on issues as varied as forest policy, coal production and utilization, water projects, and localized air pollution. Occasionally, exercises are mandated by legislation: the U.S. National Environmental Policy Act, for example, specifies opportunities for public commentary on environmental impact statements. Obviously, such exercises are not consciously designed as a way to undermine the dominant system.

Indeed, they may be undertaken with cooptation of potential trouble-makers in mind,[30] or as a veneer for decisions reached independently by conventional political means.[31] However, the very idea that these exercises revolve around multiple submissions and extended discussion means that they do not submit to the authority of expertise and established purpose. So the claim to legitimacy of incipient discursive designs is ultimately rooted not in administrative expertise or constitutional authority, but rather in ideals of free discourse (communicative ethics), in which the only legitimate power is that of the better argument.[32] Thus, if corporate or governmental actors enter such exercises and proceed to deceive, manipulate the agenda, withhold information, or establish constraining procedural rules, they may be readily exposed. Even if they are not intended as such, discursive designs can constitute "worms in the brain" that further undermine the logic of administration and the skewed distribution of power in liberal democracy.[33]

Incipient discursive designs also constitute an opening between the state and more obviously autonomous and authentic public spheres, in that the former's very existence is a concession by government authorities that professional expertise backed by state authority is insufficient to produce effective policy, and that input from a broader public is necessary.[34] In this opening, the discursive style of the public sphere might eat away at the administrative state by continually undermining claims to established, uncontroversial aims and purposes and neutral expertise. Alternatively, these exercises might be used to effectively coopt and neutralize potential trouble-makers from the public sphere. As they stand, discursive designs have an ambiguous potential in terms of constituting and facilitating political transformation.[35] Discursive exercises that are directed at policy but not sponsored by the state (e.g., the Alaska Native Review Commission)[36] have a less ambiguous character, but they are also rarer.

This ambiguity underscores the need for the construction and maintenance of *autonomous* public spheres, whose confrontation with the state is unremitting. Here, the potential of new social movements, concerning peace, community, ecology, feminism, urban space, and so forth, comes into play. Such movements may be characterized positively in terms of their self-limiting radicalism (because they do not seek a share of, much less to democratize, state power) and a perpetual concern with their own identity,[37] and negatively in terms of their unclear location in traditional class politics. The degree to which these movements should consort with established power remains controversial. Environmentalists willing to negotiate with opponents are criticized by those who prefer a more uncompromising stance. This kind of debate permeates the German Green Party, whose "realos" want compromise with conventional party politics, in contrast to the "fundis" who prefer direct action, the streets, and the public sphere. Among the U.S. greens, hostility to the processes of conventional politics has prevented most of the movement from even declaring itself a party. Clearly, "politics" and "democracy" are themselves contested concepts within new social movements, for there are internal disagreements as to just what kinds of politics and democracy are appropriate. But the very fact that such issues are on the agenda of new social movements suggests that their communicative rationality is greater than that found in conventional politics.

There is no need for premature closure of such debates in the name of unity. As Torgerson notes, uncompromising opposition to any flirtation with the state only makes sense in terms of a vision of tactical unity in the environmental movement—in other words, as an oppositional mirror image of the administrative mind.[38] Such monolithic visions, however, are as obsolete as the administrative structures they oppose.

The contemporary emerging public sphere flourishes upon diversity. And one aspect of this diversity that should be welcomed is the very refusal to consort with incipient discursive designs on the part of some environmentalists. As David Brower, the archdruid, frequently notes, one benefit of the emergence of uncompromising environmentalists such as those in Earth First! is that they make environmentalists such as himself look reasonable. So the maintenance of distance on the part of uncompromising environmentalists gives credibility to the compromisers, and makes it more likely that they will be able to effectively subvert the logic of administration and liberal democracy.

The complex relationship between discursive and liberal democracy can now be clarified. Liberal democracy's continued quest for legitimation can lead to its discursive modification (and that may be one of the strongest arguments on its behalf). But, at the same time, liberal democracy strictly limits the degree to which incipient discursive designs can affect the course of public policy, which remains subject to the accumulation imperative I discussed earlier. Within these limits, the public sphere's discursively constituted challenge to the liberal democratic state can flourish. So although simple confrontation between liberal and discursive democracy can occur, there is also a more ambiguous terrain, where the two intersect. In this terrain, the liberal democratic state can subvert and coopt discursive forms. Also, however, discursive democracy can erode the strategic, private interest-driven character of liberal democracy.

In short, the prescription that emerges here is for democratization at all possible levels: in the autonomous public spheres such as those constituted by new social movements, at the boundaries of the state where legitimacy is sought through discursive exercises, and even within the state, for example, in the form of impact assessment. This prescription parallels the thinking of theorists of participatory democracy such as Poulantzas and Gould, who, recognizing the likely persistence of the state, bureaucracy, and private enterprise, recommend democratization on a broad front, in civil society and the state.[39] Poulantzas favors this program in part to protect and build upon the worthwhile aspects of liberal democracy (such as the idea of competing centers of power), rather than sweep them aside in the name of some hoped-for mass democracy.

Let me stress here that discursive democratization is not necessarily tied to the decentralist agenda of many environmentalists. Other things being equal, small may indeed be beautiful in political and economic organization. But many environmental problems transcend the local level, and some of the more intractable ones are global in scope. Institutions of scale appropriate to deal with such issues are necessary. And there is no reason why such institutions cannot themselves be designed discursively, even though the question of unrestricted participation becomes more problematical (though not intractable) as institutional scale increases. Discursive institutional experiments can already be found in the international system, as can international public spheres, constituted by greens, peace and human rights activists, indigenous peoples, and others.[40] Pursuing democracy on as broad a front as possible is, of course, hazardous. If it involves flirting with the state and corporations, there is always the danger of cooptation and subversion (industrial democracy is particularly susceptible to this problem). And that is why one needs uncompromising public spheres to flourish and retain their autonomy.

My argument to this point has established the thesis that an ecologically inept institutional order of capitalism plus the administrative state leavened by liberal democracy is vulnerable to change in the direction of a more open and discursively democratic alternative. I have no blueprint for this kind of polity and society that would result, and indeed I would argue that the articulation of such a blueprint is undesirable—to specify it would imply a closed-endedness in institutional experimentation thoroughly at odds with the spirit of discursive designs and discursive democratization I am calling for. One of the more attractive features of such experiments is that they are open-ended and that institutional change is itself on the agenda. One sees this most clearly in new social movements, one of the

defining features of which is a permanent concern with their own
identity and power structure.[41] But any institution will exhibit such an
open-ended intelligence to the degree it embodies canons of uncoerced
communication, for there are no grounds for ruling out transformative arguments, and action based upon such arguments.

8. The Ecological Rationality of Discursive Politics

One large question remains: To what extent would the process of
discursive democratization that I have sketched mean enhanced realization of ecological values? Is it not conceivable that the individuals
involved in discursive designs would reflectively and competently
choose to downgrade environmental concerns in comparison with
(say) economic prosperity or social integration? Might they not reflect
(diminishing) cultural propensities to regard man as properly dominant over nature? In this context, it is noteworthy that Habermas
proposes types of communicative interaction embodied in discursive
designs and the public sphere that prevent interactions between persons becoming like our interactions with the natural world.[42] He
consigns the latter to the domain of instrumental-analytic rationality
and, by implication, the administrative state.[43]

Discursive democratization, however, is indeed ecologically rational, particularly when argued from the points of view of sensitivity
to feedback signals, complexity, generalizability, and compliance.

Discursive designs promote sensitivity to signs of disequilibrium
in human–nature interactions because their sine qua non of extensive
competent participation means that a wide variety of voices can be
raised on behalf of a wide variety of concerns. And to the extent that
communication is free, these voices and concerns will not be distorted
by ideology. Nor will they be moderated or distorted by the strategiz-
ing necessary under more familiar forms of liberal democratic political
interaction—where, for example, environmental interest groups
must present every issue in terms of a doomsday scenario. Nor will
they be constrained by the legal system's rules about the admissibility
and nonadmissibility of certain kinds of argument (e.g., under the
U.S. National Environmental Policy Act, the courts accept challenges
based only on the adequacy of an environmental impact statement,
not on the substantive ecological merits of a proposal).

Sensitivity to a variety of interests and issues relates directly to
the question of complexity. Ecological problems are complex, featur-
ing as they do large numbers of, and extreme variety in, the elements
and interactions facing any (human) decision-making and implemen-

tation process. Adding an unrestricted number and variety of human voices to deliberations on an issue may at first sight seem capable only of exacerbating complexity and making problems still more intractable. But this probability will apply only so long as a strategic style of political interaction obtains. To the extent participants in interactions are committed to the principles of communicative rationality, and therefore willing to renounce strategy, deception, distortion, and manipulation, then the possibility of felicitous understanding across the individuals who represent the diverse facets of complex problems becomes conceivable. The interactions between the multiple facets of a problem that define its complexity can be matched by joint, cooperative problem solving among the individuals concerned with these multiple facets. Obviously, achievement of communicative rationality can only be a matter of degree, not an absolute accomplishment. But just as one can easily identify and expose systematically distorted communication in purportedly discursive forums,[44] so one can applaud cases that more closely approximate these ideals. Perhaps the most frequently cited examples are the various inquiries presided over by Canadian justice Thomas Berger (though the inquiries are unrelated to the judicial branch). The first, sponsored (much to its eventual regret) by the Canadian government, was instrumental in blocking construction of proposed oil and gas pipelines from the Canadian Arctic to southern markets.[45] The second, constituting a more explicitly autonomous public sphere with no governmental sanction, dealt with the situation of Native Alaska in the wake of the disastrous 1971 Alaska Native Claims Settlement Act.[46] In both cases, Berger created a traveling forum for the informed articulation of the problems and needs of Northern Native peoples, providing not just a platform, but also expertise, finance, interchange, and argument. The product in each case was a coherent plan for dealing with a complex and troublesome situation.

This sort of outcome is conceivable in discursive designs, despite the formidable problems in achieving it, as well as all the usual difficulties associated with large-scale and widespread participation. This outcome is *not* conceivable in the hierarchical division of labor that defines administration, or in the strategic give-and-take of the polycentric problem solving of liberal democracy. The former only can produce a series of problem displacements, the latter can produce only self-perpetuating chaos or arbitrary compromise.

Generalizability refers to the *kinds* of values and interests that will surface in discursive interaction, and which place no limits on the kinds of interests and values that may be raised. However, interests that are generalizable to all the parties involved have much more in the way of persuasive power than interests that are particular to one or a subset of the parties. And any particular interests that are raised must survive the test of discursive scrutiny. In an ecological context, generalizable interests will often refer to public goods (in the microeconomic sense) or the quality of a common property resource such as fishery or an ecosystem. Thus the communicative interaction of discursive designs constitutes a decentralized means for supplying public goods and/or preventing the tragedy of the commons. The market has no such mechanism, and liberal democracy promotes and responds to particular, special interests, denigrating the concept of any public interest greater than the sum of unreflectively held particular interests. Administration can coercively supply public goods or regulate common property—but only at the heavy price of unresponsiveness to changing environmental signals and ineffective, uncoordinated problem solving, not to mention the undermining of democracy.

I have said little about economic organization. However, one implication for economics is quite clear: economic organization should fall under discursively democratic political control. Exactly how that might be accomplished without the heavy hand of the administrative state is a major unresolved issue, made especially problematic by the ability of capitalist and market systems to punish political decisions that impinge upon their logic of accumulation.

My arguments about the ecological rationality and political congeniality of discursive democracy apply only insofar as discourse can scrutinize and penetrate ideological and cultural schemes. Of course, such schemes are powerful. Indeed, ideology and culture may help to determine even the perceived content of ecology and environmental problems. "Nature" itself can be a social construction, looking very different to romantic poets, social Darwinists, ecofeminists, Madison Avenue executives, deep ecologists, social ecologists, and Weyerhauser. Rational discourse can scrutinize such constructions and their particular applications. There may exist no single true conception of nature to be discovered, but ideological distortions can be shattered nonetheless. Here, much depends on the increasing potential for communicative competence attendant upon modernity.[47]

In sum, the ecological politics I have sketched in this chapter is discursive and democratic, and clearly very different from the currently dominant style in political and economic life. Whether expressed in market capitalism, the administrative state, or liberal democracy, this dominant style is essentially irrational in an ecological context. The contradiction and confusion this irrationality portends create openings for a different and more ecologically felicitous style of democratic politics, which, however, will not be realized without much conscious effort and struggle.

Notes

1. For example, John S. Dryzek, *Rational Ecology: Environment and Political Economy* (Oxford: Basil Blackwell, 1987).
2. See John S. Dryzek and James P. Lester, "Alternative Views on the Environmental Problematic," in James P. Lester, ed., *Environmental Politics and Policy: Theories and Evidence* (Durham, N.C.: Duke University Press, 1989), pp. 320–322.
3. For example, Julian Simon, *The Ultimate Resource* (Princeton, N.J.: Princeton University Press, 1981), pp. 15–27.
4. Nicholas Georgescu-Roegen, "Energy Analysis and Economic Valuation," *Southern Journal of Economics*, 45, 1979.
5. Charles E. Lindblom, *Politics and Markets: The World's Political–Economic Systems* (New York: Basic Books, 1977).
6. Fred Block, "The Ruling Class Does Not Rule: Notes on the Marxist Theory of the State," *Socialist Revolution*, 33, 1977; Charles E. Lindblom, "The Market as Prison," *Journal of Politics*, 44, 1982.
7. Daniel Bell, "The Public Household: On 'Fiscal Sociology' and the Liberal Society," *Public Interest*, 37, 1974, p. 43.
8. Lindblom, *Politics and Markets*, op. cit.
9. Douglas Torgerson and Robert Paehlke, "Environmental Administration: Revising the Agenda of Theory and Practice," in Robert Paehlke and Douglas Torgerson, eds., *Managing Leviathan: Environmental Politics and the Administrative State* (Peterborough, Ontario: Broadview, 1990).
10. Garrett Hardin, "The Tragedy of the Commons," *Science*, 162, 1968; Robert L. Heilbroner, *An Inquiry into the Human Prospect: Updated and Reconsidered for the 1980s* (New York: W. W. Norton, 1980).
11. Lynton K. Caldwell, "Environmental Quality as an Administrative Problem," *Annals of the American Academy of Political and Social Science*, 400, 1974, pp. 103–115.
12. Hugh Stretton, *Capitalism, Socialism, and the Environment* (Cambridge: Cambridge University Press, 1976).
13. William P. Ophuls, *Ecology and the Politics of Scarcity* (San Francisco: W. H. Freeman, 1977).
14. Karl R. Popper, *The Open Society and Its Enemies* (London: Routledge and Kegan Paul, 1966).
15. Robert Paehlke, "Lost Keys and No Engine: Re-Starting History in the Age of Ecology," paper presented at the Workshop on Ecology, Committee on the Political Economy of the Good Society, American Political Science Association Meetings, San Francisco, 1990. See also Alex Demirović, "Ecological Crisis and the Future of Democracy," *Capitalism, Nature,*

Socialism, 1(2), no. 2, Summer 1989, p. 42, reprinted in this volume as Chapter 13.

16. Robert Paehlke, "Democracy, Bureaucracy, and Environmentalism," *Environmental Ethics, 10*, 1988, p. 305.

17. See, for example, John A. Baden and Donald Leal, eds., *The Yellowstone Primer* (San Francisco: Pacific Research Institute for Public Policy, 1990).

18. Jack A. Goldstone, ed., *Revolutions: Theoretical, Comparative and Historical Perspectives* (New York: Harcourt Brace Jovanovich, 1986), pp. 207–317.

19. James O'Connor, *The Meaning of Crisis: A Theoretical Introduction* (Oxford: Basil Blackwell, 1987), p. 148.

20. Ernesto Laclau and Chantal Mouffe, *Hegemony and Socialist Strategy: Towards a Radical Democratic Politics* (London: Verso, 1985).

21. Ibid., p. 167.

22. Jürgen Habermas, *Communication and the Evolution of Society* (Boston: Beacon Press, 1979), p. 90; John S. Dryzek, *Discursive Democracy: Politics, Policy, and Political Science* (Cambridge: Cambridge University Press, 1990), p. 17.

23. Jürgen Habermas, *Structural Transformation of the Public Sphere: An Inquiry into a Category of Bourgeois Society* (Cambridge, Mass.: MIT Press, 1989).

24. There may also be confusion in the private sector, but this is vulnerable only indirectly, through pressure on explicitly political institutions.

25. For a survey, see O'Connor, *The Meaning of Crisis*, op. cit., chap. 3.

26. James O'Connor, *The Fiscal Crisis of the State* (New York: St. Martin's Press, 1973); Claus Offe, *Contradictions of the Welfare State* (Cambridge, Mass.: MIT Press, 1984).

27. Theda Skocpol, *States and Social Revolutions* (Cambridge: Cambridge University Press, 1979).

28. For example, Lester C. Thurow, *The Zero-Sum Society: Distribution and the Possibilities for Economic Change* (New York: Basic Books, 1980).

29. John S. Dryzek, "Discursive Designs: Critical Theory and Political Institutions," *American Journal of Political Science, 31*, 1987.

30. Douglas J. Amy, *The Politics of Environmental Mediation* (New York: Columbia University Press, 1987).

31. Ray Kemp, "Planning, Public Hearings, and the Politics of Discourse," in John Forester, ed., *Critical Theory and Public Life* (Cambridge, Mass.: MIT Press, 1985).

32. Douglas Torgerson, "Limits of the Administrative Mind: The Problem of Defining Environmental Problems," in Paehlke and Torgerson, eds., *Managing Leviathan*, op. cit., p. 144.

33. Robert V. Bartlett, "Ecological Reason in Administration: Environmental Impact Assessment and Administrative Theory," in Paehlke and Torgerson, eds., *Managing Leviathan*, op. cit., p. 82. See also Michael Gismondi and Mary Richardson, "Discourse and Power in Environmental Politics: Public Hearings on a Bleached Kraft Pulp Mill in Alberta, Canada," *CNS, 2(3)*, no. 8, October 1991, reprinted in this volume as Chapter 12.

34. Torgerson, "Limits of the Administrative Mind," op. cit., p. 142.

35. Dryzek, *Discursive Democracy*, op. cit., chap. 4.

36. Thomas Berger, *Village Journey: The Report of the Alaska Native Review Commission* (New York: Hill and Wang, 1985).

37. Jean Cohen, "Strategy or Identity: New Theoretical Paradigms and Contemporary Social Movements," *Social Research, 52*, 1985.

38. Torgerson, *Limits of the Administrative Mind*, op. cit., p. 144.

39. Nicos Poulantzas, *State, Power, Socialism* (London: Verso, 1980); Carol C. Gould, *Rethinking Democracy* (Cambridge: Cambridge University Press, 1988).

40. Dryzek, *Discursive Democracy*, op. cit., chap. 5.

41. Cohen, "Strategy or Identity," op. cit.

42. Jürgen Habermas, "Reply to My Critics," in David Held and John B. Thompson, eds., *Habermas: Critical Debates* (Cambridge, Mass.: MIT Press, 1982), pp. 243–245.

43. For intimations of discursive politics relating to the biosphere in noninstrumental fashion, see Demirović, "Ecological Crisis," op. cit.; John S. Dryzek, "Green Reason: Communicative Ethics for the Biosphere," *Environmental Ethics, 12*, 1990.

44. Kemp, "Planning, Public Hearings," op. cit.

45. Thomas R. Berger, *Northern Frontier, Northern Homeland: The Report of the MacKenzie Valley Pipeline Inquiry* (Toronto: James Lorimer, 1977).

46. Berger, *Village Journey*, op. cit.

47. Jürgen Habermas, *Theory of Communicative Action* (Boston: Beacon Press, 1984).

10

Environmentalism and the Liberal State

Margaret FitzSimmons
Joseph Glaser
Roberto Monte Mor
Stephanie Pincetl
Sudhir Chella Rajan

1. Introduction

The political construction of environmentalism in the industrial democracies and in countries within their political–economic sphere raises a number of provoking questions about the confrontation of environmental movements with the liberal state and about the liberal form of institutional environmentalism that results. This chapter addresses the question of "environmentalism and the state" along a particular political and historical dimension. It lays out one of several possible lines of connection between First and Third World political struggles, asking how a particular liberal–capitalist political culture seeks to become hegemonic internationally and thus to define, structure, institutionalize, and constrain environmentalism. We intend this chapter to be a contribution to discussion of the forms of en-

Margaret FitzSimmons is Associate Professor in the Graduate School of Architecture and Urban Planning, University of California, Los Angeles. The other authors were formerly graduate students at the Graduate School of Architecture and Urban Planning. An earlier version of this chapter appeared as the "Introduction" to *CNS 2(1)*, no. 6, February 1991, pp. 1–16.

gagement and resistance that left struggles against that culture might explore.

This chapter posits the United States as paradigmatic of the liberal state and considers the institutionalization of a particular form of political environmentalism, liberal environmentalism, as the concomitant of this state paradigm. The specific political–economic history of the United States has led to a particular form of liberal institutional environmentalism; the power of the United States in the post-World War II international order has served to extend both its institutional forms of state environmental management and the discourse of U.S. environmentalism into a number of countries within its sphere of influence. Other political traditions also relate forms of environmental management through other "postcolonial" pathways between North and South; however, the hegemonic pressures of the discourse and state forms of environmentalism developed in the United States in the last century, particularly in the postwar decades, are also manifested in these other traditions.

The extension of liberal environmental forms first developed in core countries into state form and social action in the Third World gives rise to an anomaly in liberal political theory: while liberal political theory nominally privileges the autonomous nation-state, exemplars of appropriate forms of state intervention and political action themselves carry hegemonic power. Such exemplars therefore serve to extend particular compositions of social power beyond the explicit territorial bounds of their origin. Pressures for this extension of institutional form and social discourse from core to periphery come from many sources: from the dominant state itself; from economic actors such as multinational corporations that seek to universalize "rules of the game" they have successfully established in their countries of origin; and from local elites who claim power nationally and internationally on the basis of their advocacy of "modernization" and "rational public management" of popular environmental struggles.

This extension is traced here along specific lines, particularly to Japan, the Philippines, and Mexico. In Japan after World War II the United States designed and imposed a new constitution that build on the corporatist foundations of the Meiji Restoration; in the Philippines, postwar and postprotectorate governments and elites have remained strongly influenced or directed by U.S. policies and practices; and in Mexico, U.S. capital and U.S. environmental organizations now intersect indigenous agendas in the struggle over the redefinition of the Mexican polity and in debates about the North American Free Trade Agreement (NAFTA). In these and other instances, it is not only the condition of the environment but also national autonomy that is contested in alternative practices of environmentalism.

2. Where Does the Liberal State Intervene in Environmental Issues?

Struggles over the management of the environment as a reservoir of "natural resources," the necessary conditions of production, appeared relatively early in the European settlements of what would become the United States. In the absence of antecedent state forms and geographies, the institutions of the state that resulted from these struggles—first, the displacement of North American first peoples and the ensuing practices of land distribution and subsidy of infrastructure such as railroads, ports, and canals; later, the division of the remaining lands into National Forests, Bureau of Land Management lands and National Parks; and most recently, the regulation of industrial practices to contain the destructive effects of capitalist production—allowed the relatively free development of patterns of land ownership and practices of resource exploitation well coordinated with the accumulation patterns of their time.

These U.S. institutions of land management and industrial regulation have subsequently served as models for the institutionalization and regulation of environmental struggles in other industrial and industrializing countries. This transfer of institutional forms has occurred both directly, through U.S. intervention in the constitution of the political processes and structures of other countries, and indirectly, through a discourse that associates modernization, representative democracy, scientific rationality, political pluralism, and liberal economic forms and that assembles institutions that define and contain popular environmental claims. The political institutions closely related to these state forms confirm the moments of intervention they construct; the institutional environmental organizations that have arisen in the United States support both the practices and the limits of state intervention. U.S. forms of liberal environmentalism arose within a particular political economy. As this economy became international, the hegemony of its associated institutional forms has also expanded. U.S. power, and the ideological importance of images of the United States as a successful, modern, free-market democracy, have thus encouraged the widespread duplication elsewhere of environmental institutions built within the intersection of U.S. constitutional theories, political culture, capital fractions, and institutional social movements. These mirror institutions of state agency and political action form what John Dryzek terms "a nexus of capitalism, liberal democracy, and the administrative state."[1]

To consider the peculiar engagements and silences of U.S. forms of environmentalism, it is useful to contrast liberal theories of the state with structuralist theories of the liberal state. The former theories inform how institutional environmental movements within the liberal state understand their own actions and agendas; the latter theories explain how these might be more critically examined to promote understanding of political initiatives and openings. Our emphasis on the "liberal" capitalist state results from our assertion that the particular forms of U.S. environmentalism with which we are concerned arise out of the political culture of this movement as it has developed in the United States. However, the liberal state is only one possible form of capitalist state, and the spread of U.S. influences to other countries should not be taken to imply that these countries themselves have "liberal" states. For example, Japan stands as a complex example of the superimposition of particular liberal forms on a corporatist capitalist state which continues many of the economic and political institutions of the Meiji Restoration. Similarly, the Philippines and Mexico are predominantly nonliberal states, though each has adopted (or had imposed on them) certain institutions and ideologies of liberalism. Each has a complex ethnic makeup, and at a political–economic level each has strong and well-defended corporatist institutions and alliances—in the Philippines, the elites that have persisted through the series of colonial and neocolonial engagements, and in Mexico the PRI (Partido Revolucionario Institucional), a once-revolutionary universal party that has mutated to support the development of Mexican capitalism.[2] The institutional environmental movements developing within each of these countries seem, at present, to be substantially influenced by U.S. liberal forms. Such institutional environmentalism remains attached to a closely integrated discourse that associates modernization, science, progressive administrative government, and a private market economy with the resolution of environmental problems and the containment of popular environmental claims. Alternative movements, however, may offer an effective basis for change in the practice of environmental struggle, in ways that will encourage openings in the stabilized patterns of U.S. political environmentalism.

3. Liberal Conceptions of the State

To understand the points at which the state has been moved to intervene in environmental issues, it is useful to examine the liberal conception of the state and the language that supports its intervention or disengagement. This defines (ideologically) the structures and boundaries of "the political," and thus provides both internal explanation and legitimation. From this perspective, the representation is the real, in the sense that it is this representation that has been constructive of the institutional moments of U.S. environmentalism. The liberal conception of the state enables, directs, and legitimates particular forms of political action while disallowing other forms and claims. To the extent that this conception retains hegemony, it allows only certain actions spaces to appear without coercive response.

The U.S. state, with its putative balance of power between legislature, executive, and judiciary, carries within it all of the problems inherent to representational political democracy in a nondemocratic economy. It must internalize conflicting factions in order to function properly, and it must at the same time contain the political within a limited domain. Theodore Lowi's classic study, *The End of Liberalism*, anatomized the state's paralysis in setting policy whenever it must elaborately balance special interests. Only at those moments when these interests are disorganized can the state act aggressively.

One such moment occurred in the late 1960s and 1970s, when the state was moved to intervene actively in the production decisions of firms to a degree not seen before. Justification for this intervention was founded on the perceived need to contain the negative environmental effects of increasingly damaging industrial processes. The origins of this new moment of regulation have not been fully studied. It appears that the new federal initiative in environmental regulation was in some sense overdetermined, arising out of a coalescence of responses to and developments within the civil rights and environmental movements of the early 1960s and also, materially, out of the rapidly increasing environmental deterioration that occurred largely as a consequence of rapid growth of the petrochemical and defense industries in the 1950s and 1960s. Andrew Szasz[3] suggests that certain now-crucial segments of federal environmental legislation arose as a consequence of conservative attempts to displace even broader regulatory initiatives, as when the Nixon administration supported hazardous waste regulation as part of a strategy to limit federal involvement in the more general problem of solid waste disposal.

Federal intervention in the environmental question in the late 1960s and early 1970s in the United States had two conceptual phases. The first phase, from 1969 to 1977, was directed toward protection of the environmental commons (air, water, the environment as a whole); this period saw enactment of the National Environmental Policy Act, and substantial and aggressive amendments and reenactments of the Federal Clean Air and Water Pollution Control Acts.

The second phase took up the question of industrially generated environmental poisons, beginning with the Occupational Safety and Health Act in 1970 and continuing through the Resource Conservation and Recovery Act (1976), the Toxic Substances Control Act (1976), and the Comprehensive Economic Recovery and Chemical Liability Act (Superfund) (1980).[4] The legitimacy of both moments of regulation was supported (and the boundaries of the underlying conception of policy determined) by the brief intersection of a new ecological rationality of "zero discharge" and an economic rationality of the internalization of externalities, that is, of costs otherwise borne either by the public as a whole or by vulnerable workers or residents of industrial communities.

Though these two justifications for state intervention into hazardous production coincided momentarily in the early 1970s, the liberal economic model quickly regained dominance within the institutional environmental movement, as it also did in the larger policy discourse. In its acceptance of market rationality as a locus for discovering optimum public choice, this model of environmentalism conflates democracy with capitalism. Thus, in formal debate over the environmental question, both environmentalists and industry make use of a common language of economic rationality. At first, environmentalists expanded this discourse to support arguments for state intervention in those industrial processes that spilled over into the common environment: regulation was justified on the basis of market failure. Where individual actions affected the environmental commons, the state, it was argued, in the public interest, could regulate such actions by enacting laws that would incorporate external costs to correct the workings of an imperfect market.

To the extent that the liberal institutional environmental movements turned to a neoclassical language to justify their claims for environmental regulation, they won—in the short term. In the long term, however, they buttressed the legitimacy of a language that essentially represents the position of capitalist industry. This use of arguments from economic rationality to legitimate intervention into an otherwise "free" economy left an opening for the industrial counterattack, which (as early as the Ford administration) began to turn to benefit–cost analysis to justify—and, in effect, to limit—state intervention.[5] More recently, business and liberal environmental groups have agreed to extend the market to resolve (but, at a deeper level, only to disguise) arenas of environmental struggle through devices such as marketable pollution permits.

The U.S. model of environmental regulation is founded upon a concept of political life that represents society as a Hobbesian social contract among competing individuals, individuals not to be distinguished in the political arena by their wealth, power, or social form (e.g., corporations are treated as political individuals and awarded the right to free political speech). Thus, the U.S. model of political life asserts that all "interests" (the transient collectivities of liberal pluralism) have equal access to power and that democratic debate and rational decision making prevails. This model contains the political by means of a series of limits and filters that maintain the perception that economic life and private life have no politics. One aspect of this filtering is the "taboo" against direct government intervention in the production decisions of firms.[6] But, as Martin Carnoy writes,

> The primary problem of advanced capitalist societies, after two centuries of economic growth, is no longer the adequacy of resources of their "efficient" allocation for maximum output. The *way* that output is produced, the definition of what constitutes output, *what* is produced, and *who decides* development policy are the significant "economic" problems today. These problems are settled as much in the political arena as in production.[7]

Liberal environmentalism does not confront these questions about production; instead, it accepts its assigned position within the "free competition of ideas" of liberal pluralism. The source of this acceptance lies in part in the composition of the leadership of the institutional movement. Intervention consists not just of regulatory action but also of definition: of the domain of politics, of the nature of Nature, and of the origins of the environmental problem. One form of such definition arises in the selection of leaders and intellectuals to represent the movement. For the environmental movement (in comparison with other social movements), there is a special temptation here: the particular enfranchisement of expertise in the natural sciences as a seemingly politics-free domain of public choice, and from this, of technical knowledge.

The institutional environmental movement is led by scientists, neoclassical economists, and lawyers, all of whose special power depends on the continuation of the current hegemony of expert divisions of intellectual labor that support scientific and technical management of political issues.[8] For example, the leading role of the research group Resources for the Future in the discourse out of which U.S. federal environmental laws arose is clear, and this group, made up primarily of economists, was funded for years by the Ford Foundation—one of the Foundation's many contributions to furthering elite planning and management (in the sense of political control) of looming popular political issues. Whether as a result of theoretical myopia or of strategic choice, the professionals who serve the institutional environmental movement as its theoreticians accept the liberal conception of the state, and their sense of possible environmental agendas is limited by this accommodation.[9] The institutional movement remains wedded to the presentation of the state as the locus of the intersection of rationality and politics, while the interplay of politics and economics is either ignored or regarded as transparent. The liberal environmental movement, like the state to which it subscribes, disregards its own history and turns to changing technology as both the source and the solution of "the industrial problem."

4. Conceptions of the Liberal State

In a series of contributions and exchanges in *Capitalism, Nature, Socialism*, James O'Connor, Daniel Faber, John Wooding, and Charles Noble discuss a two-moment approach to understanding the U.S. state: that is, its position both as a "state in capitalist society" and as a "capitalist state."[10] In the first respect, a pluralist and liberal state is "up for grabs," presiding neutrally over the various struggles between interest groups; in the second respect, a capitalist state deeply embedded in business and politics (accumulation and legitimation) becomes trapped in a capital-oriented regulatory role.

This double formulation has its roots in a number of earlier analyses of politics and capitalism. In the 1970s there was renewed attention to critical understanding of the liberal state. Poulantzas and Aglietta in France; Habermas, Offe, and the various "state derivationists" in Germany; and O'Connor and others in the Kapitalstate group in the United States—all took up the question of the state in capitalism. These arguments reveal certain distinctive patterns. The French theorists have generally taken the centralized state for granted and its structure as given, focusing on the function of state apparatuses in managing the conditions of accumulation in relative autonomy from capital. The Germans have been more concerned with the conditions of acquisition of state power and the accumulation of state function and form in response to the specific crises, not the general logic, of capital accumulation at particular times and places. German theories are also more open to a Weberian notion of the quasi-autonomous development of the state in itself, once it has assumed certain functions.

United States work on the state is diverse and is influenced by both of these European traditions. It has tended to focus on the potential conflicts between a capitalist economy and a democratic polity, and on the problems of state provision of the minimum requirements of the social wage through taxation. But, in general, this early work tended to limit the domain of the political to struggles over the economic role of the state itself, not (in the United States instance) so much as a producer, but as an institution bound to further the conditions of general accumulation. Also, this work was connected to the social movements literature only where issues of state form and function became engaged.

John Mollenkopf[11] has developed a critique of state theories as applied to urban issues, and his work can be extended to point a direction for development of our understanding of the intersection of the state and political life. He suggests that liberal conceptions of the state treat the state simply as an arena for political struggle and thus locate the impetus for political struggles outside of the state. This reduces questions of power to interest group formation. Similarly, structuralist versions of the liberal state erase the construction of political order and emphasize economic determinism, accumulation, and regulation. But, Mollenkopf suggests, the state is neither a static structure nor a mere space for articulation of antagonistic interests; rather, it is continuously sustained, or threatened, by mobile political coalitions that are critical to its constitution. He puts forward the concept of "dominant political coalitions" as a point of departure for analysis, stressing the idea that dominant groups sustain and direct governmental power on a range of issues over an extended period. Systematic and often cross-class power is thus held by dominant groups both outside and within the state, groups (like the institutional environmental movement) that take on the ability to mobilize some constituencies while stifling others.

5. The Construction of Environmentalism by the Liberal State

The environmental question is a window that helps us to understand both the construction of the functions of environmental management within the state and the struggle over the domain of the political in everyday life. The institutional environmental movement exists within an antecedent state, constructed out of the resolution of earlier crises and conflicts of accumulation and legitimation. As a consequence, political action within the state must be roughly congruent with existing structures, functions, and compromises.

In the United States, the reformist environmental movement of the 1960s set its tactics in terms of the structures of struggle it found immediately effective in installing its agenda within the state. It did not counter the hegemony that set boundaries around political intrusion into production decisions. The state, in form and function, preceded the movement and the movement sought only to use its structured openings: legislation, pressure groups, rational public management, and litigation to command enforcement and court interpretation of ambiguous laws. Prior struggles had constructed both particular forms within the state (such as the Departments of Agriculture and of the Interior, with their divided responsibilities for natural resources and the public lands) and an institutional environmental movement that found its self-defining essence in the conception of nature as external—as Garden to Leo Marx's progressive, modern, industrial Machine.[12] In effect, the accumulating regulatory and institutional structure of the state established the conditions for further struggle.

Cross-national comparisons of state regulation of environmental questions suggest some interesting similarities and differences.[13] During the mid-1970s, in both the United States and Japan, popular movements stimulated by rapidly increasing industrial degradation of the environment challenged the state to respond.[14] Early initiatives were passed without successful opposition from industries not yet organized against this challenge. But then an almost immediate anti-regulatory backlash occurred, which led in both countries to a weakening of regulation in terms of statutory language, standard setting, and enforcement.

In the United States the initial legislation regulating toxic wastes resulted from a displacement of a regulatory movement oriented toward federal intervention in solid waste management.[15] This legislation, initially expected to be insignificant as a new federal intervention, in fact provided an armature for public claims on the state: public pressure for greater regulation. Increased stringency at the federal level intersected with growing opposition to implementation or siting at the local level, to construct what Andrew Szasz calls "policy Luddism," a two-pronged attack involving federal and local action which serves in fact to overcome the normal unwillingness of the liberal state to regulate the production decisions of firms.

Also, as Robin Bloch and Roger Keil point out in their analysis of changing air pollution regulation in Los Angeles, not all local initiatives necessarily result in improvements in effective regulation of industry or in greater democratic participation. Rather, some of these initiatives may be attempts to manage ongoing economic restructuring and to protect particular fractions of capital, and aspects of a local hegemony, at the expense of the most vulnerable members of local communities and in the guise of environmental regulation.[16] Bloch and Keil suggest that it is necessary to site analysis of particular environmental regulations in an understanding of local, regional, national, and international industrial dynamics, and to be sensitive to the need to find resolutions that consider jobs and local community well-being. Otherwise, what appear to be pluralist compromises between industry and environment can in fact be moments in the competitive restructuring of local capitals.

Ken'ichi Miyamoto, reviewing the history of postwar environmental regulation in Japan, has described how growth in heavy industry and a high pollution rate per unit produced have combined with rapid urbanization and increased mass production to cause a rapid and catastrophic destruction of the Japanese environment.[17] Though the environmental movement in reaction to this degradation won early victories—which led to an international image of Japan as an ecologically sensitive state—in fact Japanese industries mounted a rapid and successful antiregulatory response at the national level. Miyamoto, like Szasz, sees hope of resistance to this industry assault in the development of progressive local autonomies, calling for land use initiatives and public participation at the local level. But this hope remains speculative at this stage.

6. The State in Third World Countries

The struggle for hegemonic control takes different courses in Third World countries, and the state itself takes a more open and powerful role.[18] Two main differences can be pointed out.

First, the state—here meaning the politically constituted "modern" apparatus of the nation-state (democratic, despotic, one-party, or otherwise)—is directly and openly engaged in providing the conditions of production required for capitalist accumulation at levels compatible with international forces. Various dominant class fractions are often only partially integrated, so that the state itself contains the most powerfully constituted economic and political force. The state then must mediate the many struggles between dominant classes and fractions whose interests often conflict (e.g., on the economic plane, where national manufacturing industry seeks tariff protections that are challenged by exporters of raw materials or importers of goods from the core). Given the low social wage won by the limited capitalist labor force and the exclusion of large sectors of the population from the "formal" political and economic system, the state in underdeveloped capitalist countries is frequently drawn by both internal and external forces into pursuing the promotion of accumulation at any social and political cost. This cost may include high levels of popular marginalization, authoritarian control of society by means of military and police force (oppression, as opposed to hegemony), and systematic, permanent elimination of sectors of the population not able or not willing to adapt to the functional demands of capital accumulation. The state in Third World countries is particularly embedded within the contradiction of promoting accumulation and maintaining acceptable levels of political legitimacy, with regard to the management of the conditions of production, both social and natural.

Second, the state itself plays an open and direct role in Third World countries in the construction of political and cultural hegemony, leaving little room for the nominally disengaged, classical (U.S.) model of the liberal state. The fragility of this hegemony means that such states must often resort to coercive forms of political control. The multiplicity of ethnic and cultural groups, the strong counter- and alternative hegemonic resistance of social groups linked to pre- and noncapitalist political–economic forms of organization constantly challenge and renegotiate issues of hegemony within the state. Given the extents to which the dominant groups maintain close ties to core countries, internal struggles are enmeshed within the political and economic relations of domination between the center and the periphery itself.

In countries such as Mexico, Brazil, and the Philippines, U.S. economic, political, and cultural influence has been pervasive. Yet the precise forms this influence has taken differ substantially, and different geohistorical developments have led to quite different outcomes.

Antonio Contreras argues that given the high level of penetration of the Philippines by transnational forces, environmentalism serves largely to extend the hegemony of these forces.[19] Local environmental organizations, influenced by the Malthusian bias of the core institutional environmentalism on which they model their analysis, tax the marginalized poor as the source of environmental degradation. Philippine environmental law tends to parallel U.S. legislation, though this may disempower rural residents and the poor. This disempowering process is accentuated by resource-based development oriented toward international markets, by "postcolonial" and residually "feudal" social forms, and by the strong alliance between national elites and U.S. corporations. In this context, environmentalism, as a discourse, becomes a strand in the wider construction of modernity, even as in liberal settings it allows for the emergence of radical voices.

David Barkin, addressing the consequences of the ecology law enacted in Mexico in 1988, traces the complex connections between state-stimulated environmental groups and the nascent autonomous movement, tracing the contradictions the state must engage, between production and environment, between export-oriented development and the problems of legitimacy associated with international environmental pressures.[20] However, the nonmonolithic character of the state (and the weakening hegemony of the PRI) is clearly evident, as different state apparatuses are captured by different fractions of dominant classes (and eventually, by other factions of class or social groups, when and if specific alliances can be worked out). The modernization hypothesis inherent in the liberal environmental discourse implicitly supports technological interventions and technological rationality against those who argue that indigenous practices are ecologically sound.

In Brazil, the confrontation between liberal environmentalism and indigenous environmental political action is exemplified in struggles over rural land as livelihood, in which the state is deeply implicated. This reality has been brought out by a number of recent studies. For example, Roberto Monte Mor outlines the conflicts between the representation of local struggles by core liberal environmental movements—as depoliticized, spontaneous peasant struggles—and their real history in connection with a long process of rural radical action.[21] Michelle Melone describes the Amazon rubber tappers' use of international liberal environmental support to gain a political presence in the Brazilian internal struggle over the state distribution of Amazonian resources.[22] Mark McDonald reviews the military government's construction of immense hydroelectric dams and the challenge to the dams mounted by displaced small farmers, whose effective response to their displacement has changed the discourse of state responsibility in infrastructure development.[23]

These internal struggles have pushed contradictions within Third World states to new and changing (volatile) levels of compromise, reaffirming the contested character of the fragile hegemony that constitutes those states. The instability of most of these state structures and their vulnerability to both international and internal pressures attest to the fragility of this hegemony. It is in this context that the environmental question assumes a specific political and economic role, increasingly becoming a determinant factor in redirecting the state polity.

7. Distinctive Contradictions of the Third World State

Today's environmental problems increasingly imply a global perspective, given the economic and spatial integration of the contemporary world through both multinationals and markets. At the same time, we must also be aware that fundamental differences inform and determine specific manifestations of those problems in their diverse geohistorical contexts.[24] To understand the implications of the environmental question as presented in a global form, we must confront overconsumption (and overdevelopment) in industrialized countries with underdevelopment (extremes of wealth and poverty, and widespread underconsumption) in Third World countries.

On the side of overconsumption, mass production and mass consumption have been successful in appropriating and transforming available resources to produce levels of social wealth never before seen. But this wealth has been achieved at the cost of intense degradation of world resources and habitats. Contemporary environmental crises appear, in this context of overconsumption, as manifestations of imbalance between the needs of reproduction of capital and the biological and environmental conditions for the reproduction and the living species, including our own.

On the other side—that of underdevelopment—the failure to break the processes and relations of underdevelopment that have affected Third World countries has deepened conditions of poverty already existing in many nonindustrialized economies. But more particularly, the extension of state capitalism to Third World areas has, throughout the "postcolonial" period, both extended the commodification of nature and generated new contexts of exploitation and oppression. This has aggravated the extreme pressures of capitalist production and accumulation upon both human and natural resources, and has greatly exacerbated local causes and levels of environmental degradation.

The distribution of the burden of degradation of the environment, like that of any other good or bad produced under capitalism, tends to be very unequal between rich and poor. The patterns and degrees of inequality of access to environmental amenities (the "commons" of water and air, landscape, liveable spaces, etc.) have, both on the world scale and within nations and regions, sharpened dramatically in recent decades. However, capital is not entirely omnipotent, nor omnipresent. There are many areas—both geographical and social—where capitalist relations are not fully hegemonic and where political and sociocultural manifestations involve resistance to the logic of profit making. This reality raises the question of environmental struggles framed, in part at least, outside of the straitjacket of liberal political economy.

It is interesting, in these terms, to consider the circumstances of rising environmental movements in third world communities and countries. The liberal *pluralist* state, as exemplified by the United

States and the "Western" (mostly European) democracies, is not the norm in the Third World. All the same, the similar patterns of struggle associated with the dual manifestations of the state—as state-in-capitalism and as capitalist state—can be found throughout the capitalist Third World. In particular, the liberal model of environmentalism tends to be transplanted and imposed onto other cultures and territories, as a manifestation in the peripheries of hegemonic interests in the core countries themselves. However, this model when transplanted does not operate at all in the same way as it does within the "donor" countries where it is organically part of the political fabric. In the "developed" democracies the politics of lobbying, mass protest, and democratic action can be efficacious—up to a point—in obtaining compromises in such domains as industrial pollution and land conservation. But this takes place within the context of a wealthy, relatively autonomous, modern economy. By contrast, the superposition of the liberal model on a social structure of intrusive international and national capitalism constitutes a set of veneers and distortions.

First, the acute survival problems of underdeveloped countries clearly take priority in the ways that problems of environmental quality and resource depletion are framed. Second, large numbers of hungry and sick people pressure their governments to solve their problems of daily survival, even while knowing that the willingness and capability of these governments to respond to survival problems is low. Third, the state in many erstwhile colonized territories has a high profile in asserting national identity and independence. Through all these conjunctures, the state is much more intensely drawn into social and economic management, planning, and reconstruction activities than is usual in the Western "developed" democracies. This immersion in social reconstruction (including such programmes as large-scale transmigration, industrialization, and modernization) adds to the classical capitalist contradictions of accumulation and legitimacy. Under these conditions, liberal environmental groups generally act to support the legitimacy of the state, for they obtain their own legitimacy and their operating capabilities largely with reference to liberal political forms rather than through popular grassroots energies. David Barkin points out how, for example, many of the environmental groups in Mexico are organized or sponsored by the state; and Antonio Contreras observes similarly that, in the Philippines, professionals in official agencies have close ties with national and international environmental groups.[25] This exteriority of environmental intervention highlights the flaws both in the liberal model itself (as a strategy of regulation of capital) and in the assumptions present in transplanting the model across national and cultural boundaries.

8. Conclusion

In concluding, it is useful to draw out some of the differences between Third World realities and the U.S. political situation in terms of the way environmental concerns may bear on the legitimacy and transformation of the state. Third World states have not, generally speaking, shown much ability to develop solutions that resolve the fundamental contradictions in which they are embedded—even if advances in this direction can be seen in some places.[26] The absence of complete hegemonic control of the state in these societies can, however, present opportunities for the redefinition of environmental issues. The fragility of political cultures in Third World countries, caught somewhere between the process of constituting themselves in the form of modern nation states, and the subverting of this form by transnational capital, on the one hand, and ethnic and regional squabbles, on the other hand, is an important factor here. The lack of a clear political and economic hegemony makes it particularly difficult to implement coherent policies that might express compromises between the two major conflicting functions of the modern capitalist state. However, this very incoherence allows scope for social evolutions outside the rigid institutional structures and long-established hegemonies that characterize capitalist industrial societies, and particularly the United States.

In the conflict-ridden economic and political arenas of the Third World, the state may even cease to be a really existing form (e.g., in Somalia today), or may have legitimacy and influence only in very limited domains (as in many African, and some Latin American and central/south Asian territories today). State power in such situations is greatly reduced, and transformative steps are likely to be taken wherever social groupings (classes, movements) are able to build sufficiently broad alliances—to exert pressure upon the state if not bring its paralysis, collapse, or overthrow. Whether or not these transformations are deemed progressive is another matter; but movements based on the practices of radical democracy will, in this sort of context, have a clearly oppositional character. It is at this point that the environmental question, in its global and local manifestations, can arise as one major issue (among others) engaging the redefinition (and survival) of the state itself.

In U.S. environmentalism the stakes are not so extreme—not, at least, concerning the immediate survival of the constituted state. In these less dramatic circumstances, we can currently identify a shift or evolution from highly ideologized or romantic visions of preservation or conservation, to more "radical democratic" struggles at the workplace, community, household, and everyday levels.[27] Given the continuing hegemony of the liberal state, these U.S. struggles currently take a more or less oppositional form.

The acceptance of "radical democracy," involving challenge to (rather than acquiescence to) the pluralism of representative democracy, implies recognition of the multifarious and plural character of social struggles, moving beyond the simple centrality of the locus of production in radical politics. This implies that the growth of democratic struggles in broader arenas of everyday life—such as environment, household, community, city/urban—be added to workplace and traditional politics. It is not a question of dismissing or substituting for these struggles (taken to be ontologically central), but instead a matter of supplementing by expanding the social and political space within which the struggle against capitalism occurs.[28]

Effective resistance to capitalism means dislodging the state from its role in support of internal and external capitalist accumulation, rather than dislodging the state itself. Yet certain commonalities exist in both periphery and core. In both, it is necessary to appraise, for their limitations as well as for their possibilities, the various institutions implicated in the hegemony of liberal environmentalism in core and periphery: representative democracy at a level that too often disenfranchises the locality as the "life space"; normal science and its abstracted construction of a totalizing discourse of risk and benefit; the openings or closures revealed within the administrative activities of the state; and, most crucially, the distinction between the economic and the political that constitutes the conventional boundary of the liberal state. The question is: What possibilities of redefinition of the struggle for democratization are opened by environmental struggles in both their straitjacketed (liberal, plural) and more autonomous forms? We can learn much about a more radical democracy from these challenges to the limited discourse of the liberal state.

Notes

1. John S. Dryzek, "Ecology and Discursive Democracy: Beyond Liberal Capitalism and the Administrative State," *CNS*, *3(2)*, no. 10, June 1992, p. 18, reprinted in this volume as Chapter 9.

2. See James D. Cockcroft, *Mexico: Class Formation, Capital Accumulation and the State* (New York: Monthly Review, 1983).

3. Andrew Szasz, "In Praise of Policy Luddism: Strategic Lessons from the Hazardous Waste Wars," *CNS*, *2(1)*, no. 6, February 1991, pp. 17–43.

4. For a discussion of this legislation by someone who was a senate staff member during its early enactment, see Walter E. Westman, *Ecology, Impact Assessment, and Environmental Planning* (New York: John Wiley, 1985);

see also his "Some Basic Issues in Water Pollution Control Legislation," *American Scientist*, *60*, 1972, pp. 767–773. Other industry-specific legislation regulated pesticides and consumer products, often by more stringent reenactments of existing legislation.

5. Richard N. L. Andrews, "Economics and Environmental Decisions, Past and Present," in V. Kerry Smith, ed., *Environmental Policy under Reagan's Executive Order* (Chapel Hill: University of North Carolina, 1984).

6. Szasz, "In Praise of Policy Luddism," op. cit.

7. Martin Carnoy, *The State and Political Theory* (Princeton, N.J.: Princeton University Press, 1984), p. 3.

8. This is true of the institutional movement both within and without the state *sensu strictu*; the forms of liberal environmentalism, and its intellectual leadership, present a particularly clear example of Poulantzas's analysis of the role of intellectuals in confirming the technical power of the state to govern; see Nicos Poulantzas, *State, Power, Socialism* (London: New Left Books, 1980).

9. Margaret FitzSimmons and Robert Gottlieb, "A New Environmental Politics," in M. Davis and M. Sprinker, eds., *Reshaping the U.S. Left: Popular Struggles in the 1980s* (London: Verso, 1988).

10. James O'Connor, "Capitalism, Nature, Socialism: A Theoretical Introduction," *CNS*, *1(1)*, no. 1, 1988; Daniel Faber and James O'Connor, "The Struggle for Nature: Environmental Crises and the Crisis of Environmentalism in the United States," *CNS*, *1(2)*, no. 2, Summer 1989; Charles Noble and John Wooding, "The Struggle for Nature: Replies," with Faber and O'Connor response, *CNS 1(3)*, no. 3, November 1989.

11. John Mollenkopf, "Who (or What) Runs Cities, and How?," *Sociological Forum*, *4*, no. 1, 1989.

12. Leo Marx, *The Machine in the Garden: Technology and the Pastoral Ideal in America* (Oxford: Oxford University Press, 1964).

13. The literature that contrasts state responses to environmental questions across the industrial democracies is limited, and published critical work is largely absent. However, Ronald Brickman, Sheila Jasanoff, and Thomas Ilgen, in *Controlling Chemicals: The Politics of Regulation in Europe and the United States* (Ithaca, N.Y.: Cornell University Press, 1985), provide extensive material which, though not critical in its analysis, is informative as to the different practices of environmental regulation in the United States, the United Kingdom, France, and Germany.

14. Ken'ichi Miyamoto, "Japanese Environmental Politics since World War II," *CNS*, *2(2)*, no. 7, June 1991.

15. Szasz, "In Praise of Policy Luddism," op. cit.

16. Robin Bloch and Roger Keil, "Planning for a Fragrant Future: Air Pollution Control, Restructuring, and Popular Initiatives in Los Angeles," *CNS*, *2(1)*, no. 6, February 1991, pp. 44–65.

17. Miyamoto, "Japanese Environmental Politics," op. cit.

18. We do not use the term "Third World" as a category, a place, nor even a sociospatial reality. Instead, it is used as an attribute, as characteristics. In that sense, we can recognize Third World countries, Third World cultures, Third World groups constituted with and through their localities and histo-

ries. In other words, our approach does not take Third World to be an entity, an identity, a noun, but a quality, an adjective.

19. Antonio Contreras, "The Political Economy of State Environmentalism in the Philippines," *CNS*, *2(1)*, no. 6, February 1991, pp. 66–84.

20. David Barkin, "State Control of the Environment: Politics and Degradation in Mexico," *CNS*, *2(1)*, no. 6, February 1991, pp. 86–108.

21. Roberto Monte Mor, "Incident at Bela Vista Farm, Southern Para, Amazonia," *CNS*, *2(2)*, no. 7, June 1991, pp. 22–26.

22. Michelle A. Melone, "The Struggle for the Seringueiros: Environmental Action in the Amazon," in John Friedmann and Haripriya Rangan, eds., *In Defense of Livelihood: Comparative Studies on Environmental Action*, (West Hartford, Conn.: Kumarian Press, 1993).

23. Mark D. McDonald, "Dams, Displacement, and Development: A Resistance Movement in Southern Brazil," in John Friedmann and Haripriya Rangan, eds., *In Defense of Livelihood*, op. cit.

24. Ramachandra Guha's confrontation with the American wilderness preservation movement's attempts to intervene in land control questions in the third world presents this clearly; see Ramachandra Guha, "Radical American Environmentalism and Wilderness Preservation: A Third World Critique," *Environmental Ethics*, 11, 1989.

25. Barkin, "State Control," op. cit.; Contreras, "Political Economy," op. cit.

26. For comparative examples from Latin America, Africa, and South Asia, see case studies in John Friedmann and Haripriya Rangan, eds., *In Defense of Livelihood*, op. cit.

27. Compare the positions taken by Ernesto Laclau and Chantal Mouffe in *Hegemony and Socialist Strategy* (London: Verso, 1985) with the various arguments of Murray Bookchin, associating the renaissance of civil society with the rise of an environmental consciousness which forces the recognition of community. But it is important to distinguish between theories of the present and the future, between theoretical engagements that address the actual structuring effects of capitalism and those that take up the question of the potential forms of socialist politics. Raymond Williams brought this to our attention quite clearly, in his discussion of alternative and oppositional social movements and the need for their integration, see Raymond Williams, *Marxism and Literature* (Oxford: Oxford University Press, 1977).

28. We mean here to defer to both James O'Connor's and Norman Geras's critiques of the limitations of Laclau and Mouffe's arguments and its self-conscious post-Marxism. And yet we argue that any examination of environmentalism must address its potential expansion of the political; see O'Connor, "Capitalism, Nature, Socialism," op. cit; Norman Geras, *Discourses of Extremity* (London: Verso, 1990).

11

Environmental Security and State Legitimacy

Colin Hay

1. Introduction

Over the past 20 years our knowledge and understanding of the developing ecological crisis has advanced dramatically, as have societal perceptions of this environmental degradation. Yet advanced, liberal, democratic states dependent upon societal legitimation for their very continuity have consistently failed to respond to the environmental crisis in anything other than a token fashion. In fact, as this chapter seeks to demonstrate, governing institutions have developed a complex repertoire of *responsibility- and crisis-displacement strategies*, thereby preserving their legitimacy with respect to environmental issues while in fact failing to engage in the collective, interstate responses necessary to address crisis tendencies that are inherently global in nature.

It is therefore crucial that we divert some of our attention from honing our understanding of the specific mechanisms that generate environmental crisis, and begin to consider the nature of the *political* filtering and processing *through the state* of perceptions of environmental crisis and perceived responsibility. It is my argument in this chapter that the state-centered processing of perceived responsibility for envi-

Colin Hay has taught social and political science at Cambridge University, and is currently Research Fellow and Associate Lecturer in the Department of Sociology at Lancaster University, the United Kingdom. He is a member of the British Editorial Group of *Capitalism, Nature, Socialism*. He would like to thank Wayne Hope, Bob Jessop, Martin O'Connor, Andrew Sayer, Caroline Schwaller, and the Boston Editorial Collective of *Capitalism, Nature, Socialism* for their extremely helpful comments on an earlier draft of this chapter. First published in *CNS, 5(1)*, no. 17, March 1994, pp. 83–97.

ronmental degradation represents the single major factor preventing
a global response to a global crisis. Thus, the environmental movement does not merely require a theory of ecological crisis; it requires
a theory of the state.

This chapter seeks to contribute to such a theory by developing
a heuristic framework for assessing the likely response of advanced
liberal democracies to perceived and actual environmental stress. Such
state-specific responses, however, must be located within the *economic*
context of global capitalist accumulation (within which different economies find different *modes of insertion*), and within the *political* context
generated, on the one hand, by the growth of new transnational social
movements, and, on the other hand, by the potential offered by
"global environmental diplomacy" (perhaps most clearly developed
in response to ozone depletion, and exemplified in the Montreal Protocol and Rio Earth Summit of 1992).

Since, overall, the myriad environmental crises are themselves
induced by the growth imperative of *globalized* capitalist accumulation
(albeit processed in specific ways in specific national and local contexts), concerted interstate action is essential to avert terminal global
ecological catastrophe. Yet, while in the long term it is in the collective
interests of all states to engage in concerted responses to environmental degradation, the different modes of insertion of national (and
indeed regional and local) economies within such global economic
dynamics generate a multitude of conflicting state interests that clearly
militate against the emergence of such collective responses to environmental crisis. My purpose here is to consider whether the, *state-centered*
institutional and representational logic of advanced liberal capitalist
democracies precludes the degree of concerted interstate response
necessary to address and resolve such crisis tendencies.

2. Toward a Critical Theory of Environmentally Induced Legitimation Crises

Following the analysis developed by Habermas in *Legitimation Crisis*,[1]
we can distinguish between "systems crises" ("economic" and "rationality" crises) and "identity crises" ("legitimation" and "motivation"
crises). The former refer to breakdowns in system integration, such
as periodic crises of overproduction within advanced capitalist economies due to a lack of coordination between production and consumption, and are thus structural crises.[2] The latter refer to crises involving
a general breakdown of social integration, *subjectively perceived* crises

brought about by the collective perception within civil society of the existence of structural or deep-seated contradictions.[3]

Habermas argues that the state must intervene within the (national) economy[4] to secure the conditions for continued capitalist accumulation. Such intervention is complicated, however, by the simultaneous and continual need of the state to secure and resecure its legitimacy within civil society. It is not difficult to see how this analytical framework might be extended to include the state's responsibility for environmental regulation—an extension of economic intervention necessitated by the exigencies of state legitimacy: namely, maintaining viable conditions for capitalist accumulation, and minimizing environmental degradation.[5] Similarly, it is not difficult to see the potential for tension that this problem generates for the state in seeking to reconcile the conflicting *short-term* interests of capitalist accumulation and long-term considerations of environmental preservation.

Environmental regulation, as an extension of state economic intervention, may thus be conceived, following Habermas, in terms of the *logic of crisis displacement*. Through the operation of this logic, potentially fundamental economic and environmental–economic crises (those that are grave enough to threaten the very existence of capitalism itself) become displaced to the realm of political responsibility. Here they become expressed as less fundamental political, or "rationality" crises (less fundamental because they are no longer crisis *of* capitalism, but rather political crises *within* capitalism).[6] This mode of crisis displacement characteristic of advanced capitalist economies can be seen as functional for short-term social and system reproduction, since deep-seated structural crises become articulated as crises of a particular and transient political rationality as opposed to crises of capitalist accumulation per se.[7]

Having demonstrated the displacement of economic (and, by implication, environmental–economic) crises into the political arena, Habermas proceeds to demonstrate that the existence of a political, system, or "rationality" crisis of the state is likely to precipitate a *subjective* or *perceived* "identity crisis" of the state, within civil society. Here the very legitimacy of the particular form of the capitalist state (e.g., welfare capitalism) becomes threatened as a result of the real and material experience within civil society of the *symptoms* of a rationality crisis, and of the developing perception within the public sphere of the existence of a "system" crisis within the political realm. Although such a crisis of legitimacy becomes articulated at the political level, it ultimately has its origins, Habermas argues, within the structural nexus of capitalist accumulation. Similarly, although an environmentally induced legitimation crisis becomes articulated at the political level, and becomes politically processed, it too has its origins within the fundamental economic–environmental contradiction of advanced capitalism's growth imperative.[8]

Habermas's theoretical framework, however, is not without its problems. His perspective is closely tied to the crisis of Western capitalism of the 1970s in the midst of which it was formulated. As a consequence, he fails to take sufficient account both of the state's ability to respond to crises so as to resecure its basis of legitimacy and of the variety of crisis displacement *strategies* that the state can deploy in so doing.

For example, the increased openness of the global economy undermines the demand-side strategies of the Keynesian Welfare State.[9] This enables two potential crisis displacement strategies: *either* the attempted displacement to economic actors of responsibility for economically induced environmental degradation (the strategy pursued by the Reagan administration, for instance, through environmental deregulation—effectively a governmental washing of the hands of responsibility for environamental pollution), *or* a more interventionist strategy capitalizing upon the supply-side advantages of investment in putatively environmentally sustainable niche-markets (a strategy pursued at various times by both the German and the Japanese states). Despite these qualifications, however, what remains of crucial importance to any understanding of the state's ability to respond to threats to environmental security is Habermas's recognition of the importance of both structural characteristics and subjective perceptions of crisis.[10]

3. Environmental Crisis and State Legitimacy

By constructively reworking Habermas's account of the logic of crisis displacement, we can posit the idea that specific manifestations of the fundamental and global environmental–economic contradiction, though precipitated initially within the global economic system, become displaced to the level of the nation-state and are articulated as rationality crises of the political system—that is, as political crises of environmental regulation. The state's inability to respond to the root causes of such crises is likely to become the focus of political mobilization within civil society. In such circumstances, environmental pressure groups express individuals' experiences in terms of living through a crisis scenario, and thereby find *resonance*, potentially inducing a crisis of political legitimacy.

At this stage the potential options facing the political system of an advanced capitalist state can be arrayed *hypothetically* on a scale between two extremes. The first is to enact a systematic and wholesale reorganization of the economic system (its growth imperative, mode of regulation and the relations of production it embodies), such that it is possible to overcome the incompatibility between the political rationality of state economic regulation and the requirements of environmental sustainability. The second option is to do nothing. In practice, neither strategy is feasible. For the former option would require a fundamental restructuring of the state's decisive economic nucleus to the degree that it could no longer be described as capitalist. Such a strategy could simply not be sustained within a global capitalist world economic dynamic. Similarly, the latter option would involve a profound threat to continued political legitimacy, and ultimately a profound threat to the consensually sustained stability of the state.

In practice, then, the state's actual range of options is far more limited: on the one hand, by the insertion of the national economy within global economic dynamics premised upon the growth imperative; and, on the other hand, by the necessity of maintaining continued state legitimacy (in turn dependent upon some responsiveness to the health, environmental, and security concerns of its citizens). Any state-sponsored strategy for fundamental structural change to remove the growth imperative would result in an equally destabilizing crisis involving a radically constrained economy within an unchanged global economic dynamic, or would be reliant upon an unprecedented concerted interstate response. Similarly, within an advanced democratic mode of political representation, a political regime will never fail to make some response to a fundamental crisis of the very legitimacy upon which its continued existence is premised. The crucial point, however, is that states that are reliant upon societal consent respond at a largely tactical or a cosmetic level to threats to their legitimacy, and thus respond to *subjective perceptions of crisis rather than to the contradictions and discontinuities that precipitate such threats to legitimacy.* Thus, the underlying rationality of democratic representation encourages states to restrict their responses to such crises to the minimum they perceive necessary for short-term restoration of legitimacy. This is likely to be achieved through a combination of symptom amelioration, token gesturism, the "greening" of legitimating political ideology,[11] and the displacement of the crisis in a variety of different directions: either downward into civil society (by making individuals responsible on a personal level for the response to environmental crisis through facilitation of "green" consumer choice); or upward onto a global political agenda; or, indeed, sideways in presenting the crisis as another body's (e.g., state's) legitimation problem.

As a consequence of any such minimal and *state-specific* response to *globalized* environmental risk, the crisis (the totality of all environmental contradictions) inevitably deepens, subsequently becoming

rearticulated as a further and more fundamental environmentally induced *rationality and ultimately legitimation crisis of the state*. And once
again the representational logic of the state structurally selects for a
strategy of attempting to rescue legitimacy at minimum short-term
cost through minor superstructural tinkering. Indeed, the national-
level conjunctural response will only give way to the concerted
global[12] response necessary to address the causes of such profound
environmental crisis when two conditions are met:

1. The state can no longer secure its legitimacy base through
   strategies of scapegoating, crisis displacement, and token gesturing.
2. The perceived costs of such a direct response to a profound
   crisis scenario are deemed to be less destabilizing than the
   coresponding loss of state legitimacy produced by a continued
   failure to make such a response.[13]

To take two examples, the respective problems of ozone depletion (with a perceived high-consequence risk and a low-cost response)
and global warming (with a high-consequence risk and a high-cost
response) lie at opposite ends of this continuum. It is therefore not
surprising that ozone diplomacy and the Montreal Protocol are held
up as the supreme example of concerted global action to avert ecological damage. The likelihood is, however, that we shall have no similar
global protocol or response to such issues as global warming, deforestation, or third world environmental degradation. This was indeed
demonstrated by the superficiality of the measures agreed to at the
recent Rio Earth Summit: the token gesturing of the United States
diluted treaty to curb "greenhouse gas" emissions; the lack of any
commitment to halt deforestation; and the advanced industrial nations' failure to pledge even 5% of the U.N.'s conservative estimate
for merely beginning to address third world environmental crisis.[14]

4. Strategies for Environmental Sustainability: Strong States and Weak States

Thus far I have been concerned with formulating an analysis of the
generalized constraints placed upon the state's ability to respond to threats
to environmental security, due to the state's location within a dynamic
global capitalist economy; to the representational logic of advanced liberal democracies; and to the associated need to maintain societal legitimation. I have so far failed to consider whether different strategies for

environmental crisis management and displacement are available to states by virtue of their relative "strength" or "weakness."[15] This might have some impact on how the generalized problems of environmental crisis resolution within a capitalist world economy become articulated within the policies pursued by real states at specific moments in time.

All states are "weak" in relation to threats to environmental security that can be resolved only through *concerted global* intervention designed to fundamentally restructure the internal logic of the capitalist world economy. Nonetheless, while all states are weak in this sense, some states are weaker than others. Less weak states (those with generally small, highly skilled, and flexible advanced industrial economies) can deploy strategies of state-led economic restructuring through competitive niche-marketing strategies, such that their restructured economies (or sectors within them) remain complementary to the global economic dynamic, but nonetheless *appear* environmentally sustainable. Hypothetical examples of such strategies, which are only available to a very few national economies, include the attempt to develop an "environmental protection" economy or sector within the economy (selling environmental "mopping-up" sevices to other economies),[16] and the development of a banking economy. Similar strategies pursued by real states include Japan's response to the "Oil Shock" through massive investment in energy-efficient capital, the rewards of which are still being reaped,[17] and the competitive advantage that the German economy has won through government support for environmentally clean technological research through the BMFT (Bundesministerium für Forschung und Technologie)[18] and the subsequent patenting of high technology "green" consumer items such as catalytic converters.[19] The crucial point, however, is that such examples of *apparent* localized (and occasionally nationalized) environmentally sustainable economic growth are *parasitic* upon the global capitalist growth imperative, which is itself the very antithesis of environmentally sustainable economic growth. This, then, confirms the thesis that regardless of how relatively "strong" some states might appear, in relation to environmental issues, all states are fundamentally "weak" by virtue of the capitalist growth imperative. For the latter dictates that successful economies adopt a complementary and privileged growth dynamic within the emerging anarchic dynamic of a global capitalist economy.[20]

5. Subjective Perceptions of Environmental Crisis

The theoretical framework I have developed thus far stresses the importance of legitimation and thus of *subjective* perceptions of crisis, as opposed to objective, demonstrable structural or system crises. It is not merely corporate actors, however, who are reluctant to engage in the type of radical restructuring necessary to secure environmental security. Much of the ability of political regimes to resecure, at least temporarily, their legitimacy in response to crisis derives from the short-term interests of many sections of civil society in maintaining the status quo. Because of this parochial conservatism within civil society, a *preemptive* radical restructuring of the political mode of economic regulation in response to a developing environmental contradiction is likely to prove more politically destabilizing and delegitimizing in the short term than a failure to respond. Similarly, it is likely that the changes in life-style that would be necessary to prevent continued and accelerating environmental damage would involve considerable sacrifice for most sectors of society, and an associated profound redefinition of societal "common sense." This is not to suggest that the rewards of wasteful consumption are not illusory. Rather, it is to insist that the painful reality of exponential environmental degradation necessitates far more than the "greening" of life-style identities and consumer choice. Instead, what is required is a much more deep-seated transformation of the context within which such choices are exercised. In order to restructure the social formation to this extent, an equally radical restructuring of the individual's perception of what is socially, politically, and culturally desirable would be required. For popular aspirations within civil society to be so fundamentally transformed, the nature of the environmental–economic rationality crisis would have to be extremely severe, to the extent that individuals' daily lived experiences became *unequivocally and deeply* affected by the symptoms of environmental crisis. Only such an experientially induced crisis of ontological security would be *likely* to precipitate a crisis of legitimation of the political system sufficiently deep to result in a concerted interstate response.

The environmental movement must come to terms with the causes and consequences of this logic of crisis displacement if it is to mobilize to delegitimize the states that perpetuate it. It is clear that we cannot merely wait for the development of a crisis that is experientially induced. Rather, we must give expression to this worst-case scenario in order to catalyze awareness of both environmental degradation *and* state-specific strategies of crisis displacement.

6. Typology of Environmental Crisis Tendencies

If the environmental–economic contradiction within advanced capitalism represents the totality of contemporary environmental problems, then particular environmental risks such as ozone depletion, deforestation, global warming, or species loss can each be seen as specific symptoms of this fundamental contradiction.

In assessing the likely effects of environmental degradation, and the response of states to environmentally induced legitimation crises, it is important to consider the nature of each specific problem or crisis, and thus to develop some sort of crisis-risk typology. It is, of course, essential to hold in mind a "cumulative" or synergic view. For the environmental–economic contradiction must be addressed in a concerted fashion if environmental pathology is to be averted. However, the nature of political representation within contemporary capitalism is such that states respond mainly to specific crises, not to generic crisis tendencies. Thus the form of the state's response will vary according to the perceived crisis type.

Within the heuristic Habermasian theoretical perspective outlined above, we can begin to formulate a typology of environmental crises reflecting the *state's own logic of risk assessment* as opposed to any scientific discourse about the "actual" threat posed by any particular symptom of environmental degradation (see Table 1.1). For states do not respond to crisis tendencies directly, but rather to how they impinge upon the state.[21] The state is thus engaged in a constant risk calculus that acts as a mediator between perceived environmental risks and state responses. The relative importance of these environmental risks to the state can be viewed along two axes:

1. How much they will affect routinized human experience—the level of consequence of the risk. Here we may draw insight from Giddens's notion of *"high-consequence risks."*[22]

TABLE 11.1. A Typology of Environmental–Economic Risks

Economic cost of crisis resolution	Potential damage to natural environment— Level of consequence of risk		
	High	Intermediate	Low
High	Global warming Deforestation	Mass species extinction	Destruction of rural landscapes
Medium	Nuclear disaster (industrial/ military)	Air/ocean pollution Acid rain	River pollution
Low	Ozone depletion	—	Species extinction Footpath erosion

2. The "cost" to the state of risk amelioration—the extent to which solutions would necessitate significant restructuring modes of capital accumulation.

Global warming ranks high on both of these axes—it is a high-cost, high-consequence risk—since its effect will severely threaten human activity in its current form, while its solution would require a serious restructuring of industrial production profiles, modes of transport, and so on. The problems posed by chlorofluorocarbons (CFCs) and ozone depletion (even if we temporarily omit the contribution of CFCs to global warming)[23] rank high on the first criterion (representing a high-consequence risk), and relatively low on the second (necessitating merely a low-cost response—a switch from CFCs to other aerosol propellants).[24] The threat of extinction to a particular plant or animal will, typically, rank low on both criteria.

This typology of environmental–economic risks is not, however, exhaustive. Often environmental polluters do not directly suffer the consequence of their pollution. This can be illustrated through the example of the problems posed by acid rain in northern Europe. For Britain, a chief polluter, acid rain represents a moderate- to high-cost, low-consequence risk since the consequences of acid rain are primarily felt outside Britain; yet, from the perspective of a Scandinavian government, acid rain represents a moderate- to high-consequence risk, but a problem potentially of minimum cost if the polluters are to be held responsible. Thus, a further series of categories could be introduced into the typology: risks of high/low consequence to other states, yet low/high consequence to the environmental polluter.

What this further stresses, however, is that due to the nature of contemporary political legitimation, there is no universal global environmental risk typology, and that the response of individual states to environmental pathologies is often determined by contingent political factors as opposed to informed risk assessments. The challenge that this reality poses for the environmental movement is not only to continue to provide such informed risk assessments, but also to expose the distortions imposed by the state's own consequence–risk calculus.

Similarly, it might be objected that the above typology fails to take account of the obvious links between many of the risks specified. Deforestation and ozone depletion, for instance, are integrally tied to global warming. However, in many respects this sharp demarcation of specific risk tendencies is crucial to our understanding of states' political risk assessment. For states respond to perceived crises and are also, at least in part, responsible for mobilizing or channeling such perceptions. In this way, interrelated contradictions, risks, and crises that originate out of a single and fundamental environmental–economic contradiction are articulated and responded to as specific and self-contained. The above risk typology therefore represents the state's logic of risk assessment rather than any more objective representation of crisis tendencies.

7. Conclusions: The Problems of Environmental Crisis Resolution

The aim of this chapter has been to begin to develop a heuristic theoretical perspective capable of assessing the likely responses of advanced Western states to perceived and objective environmental contradictions. The motive in formulating such a theoretically informed analysis has been to draw attention to the constraints imposed upon responses to environmental crisis by the state-specific nature of political legitimation and by the representational logic of advanced liberal capitalist democracies.

The implications of such a consideration are profound. The logic of global crisis management (as opposed to resolution) within a disaggregated world system of bounded nation-states is one of displacement of crises from the global level (the level at which they must be addressed if they are to be resolved) to the state level (a level marked by contingent and parochial political responses). The state-specific nature of crisis management within the advanced capitalist world economy thus represents a fundamental threat to continued environmental security.

The logic of state-specific responses to global threats enables a series of crisis displacement strategies and imposes a series of constraints that together conspire to generate disfunctional long-term tendencies:

- The displacement of environmentally induced legitimation crises into civil society in order to resecure short-term political legitimacy.
- The response of states to such crises so as to do the minimum perceived necessary to (re)secure legitimation.
- The parochial nature of responses to specific crises which are mere symptoms of an unfolding fundamental environmental–economic contradiction, that is, symptom management as opposed to crisis resolution.

- The political calculation of risk scenarios in terms of cost and consequence to the state.
- The logic of token gesturism, scapegoating, crisis displacement, and the ideological overrepresentation of the extent of "green" conversion.
- The response of state to subjective perceptions of crisis as opposed to the actual conditions giving rise to such experiences and perceptions of crisis.
- The state-specific nature of responses to crises and long-term contradictions that are global in origin and nature.
- The operational rationality of short-term strategic calculation as opposed to long-term preemptive and preventative action.
- The constraints imposed by a global capitalist growth imperative with which national economies must remain complementary.

Only if each of these structural political constraints to interstate action can be exposed and overcome through collective political mobilization might the necessarily concerted response to global environmental pathology be taken. Whether such a transition from a state-specific to a politically enabled global mode of crisis management and resolution will occur prior to the symptoms of terminal global environmental crisis must remain disturbingly unclear. Yet what does become clear is that the political structures of advanced capitalist states represent a profound threat to global security and stability. The challenge that now faces the environmental movement is monumental. Not only must it continue to develop and disseminate informed risk assessments, highlighting the developmental logic of environmental crisis amplification, but it must also expose the distortions imposed by the state's own consequence–risk calculus.

Notes

1. Jürgen Habermas, *Legitimation Crisis* (London: Heinemann, 1975).
2. It might be argued that within a liberal democracy a contradiction only becomes a crisis when it is perceived as such. As a consequence the term "system crisis" might be considered a redundant theoretical abstraction. While I have a certain sympathy with this view, what Habermas's distinction does allow for is a consideration of the structural properties of, and subjective perceptions of, a crisis, and the *nonnecessity of their correspondence*. This latter point, however, is sadly underdeveloped in Habermas's account of crisis dynamics, which remains primarily systems theoretical.
3. It is important here to note that such a crisis is not "merely subjective," but refers to the lived and perceived nature of a crisis. Hence, Habermas

is here seeking to distinguish between crisis as system dysfunction (contradiction) and crisis as a lived experience involving a perception of such contradiction.

4. A fundamental weakness of Habermas's approach is his failure to contextualize the national economy within global economic dynamics, and to develop an analysis of the "integral economy" more generally. See Bob Jessop, "Integrale Ökonomie und Integrale Politik: Regulation und Staat," in Alex Demirović et al., eds., *Regulation Staat und Hegemonie* (Munster: Westfälisches Dampfboot, 1991).

5. Thus, for instance, the state must regulate the use of natural resources within a national context if capitalist accumulation is to be able to continue (e.g., by promoting sustainable development). Similarly, the state must restrict industrial emissions to a certain extent in order to preserve the possibility of continued accumulation—there is a limit to how much surplus value can be extracted from a dead work force. Hence, the state must seek to preserve the basic conditions within which capitalist accumulation can proceed, and this requires a certain degree of environmental regulation.

6. At this stage, however, they still remain crises of the system as opposed to perceived or identity crises.

7. For an interesting parallel argument, see Karl Marx, "Critical Marginal Notes on the Article by a Prussian," in Karl Marx and Frederick Engels, *Collected Works*, vol. 3 (1844; London: Lawrence and Wishart, 1975), pp. 189–206

8. This is a point perceptivley made by Altvater in his pioneering study, *Zukunft des Marktes* (Munster: Westfälisches Dampfboot, 1991). Although it might be argued that it is not so much the *capitalist* growth imperative that is environmentally unsustainable as economic growth imperatives per se, in considering the contemporary environmental crisis we must consider the specificity of today's growth imperative, which is inherently capitalist. Thus, although in formulating visions of an alternative environmentally sustainable society we must be wary of growth imperatives altogether, and we must understand the contemporary environmental crisis as a consequence not of growth imperatives in abstraction, but of the *capitalist* inflexion to the growth imperative.

9. See Bob Jessop, "Towards a Schumpeterian Workfare State? Preliminary Remarks on Post-Fordist Political Economy," *Studies in Political Economy*, 40, 1993.

10. By "environmental security" I mean a conjuncture of mutual sustainability and complementarity between the natural environment, the created environment, and human communities, such that the dynamic of societal organization and/or the created environment does not challenge or threaten continued environmental preservation. The term "environmental preservation" can be specified at a multitude of different heuristic levels, from that of the species, to that of the local ecosystem, and ultimately to that of life on earth itself.

11. See, for example, Susan Owens, "Environmental Politics in Britain: New Paradigm or Placebo?," *Area*, 18, no. 3, September 1986; Mike Rob-

inson, *The Greening of British Party Politics* (Manchester: Manchester University Press, 1992).

12. "Global" in the sense of having to address, at cause, the fundamental environmental–economic contradiction of the growth imperative of the *global* capitalist economy.

13. Yet even when such conditions are met an effective response to environmental crisis is clearly not guaranteed. A concerted willingness to engage in collective interstate action to address the perceived root causes of environmental degradation does not easily translate into concerted and coherent action. The different and specific interests of individual states, and indeed nonstate players (nongovernmental organizations, transnational corporations, and grassroots popular movements) will ensure that there will be struggle. Further, this struggle will inevitably be played out on a strategic terrain that favors the interests of capital and globally dominant political alliances. Such dominant interests may very well subvert the desire to forge a concerted response to the environmental crisis, resulting either in inaction or ineffective action. Although not in any sense a concerted attempt to address the root causes of environmental degradation, the Rio Earth Summit nicely demonstrated both tendencies.

14. Donna Lascelles and Christina Lamb, "A Game of Missed Opportunities," *Financial Times*, June 15, 1992; *Guardian*, June 15, 1992; *Observer*, June 14, 1992; for a more optimistic view of international environmental possibilities, see Jim MacNeill et al., *Beyond Interdependence: The Meshing of the World's Economy and the Earth's Ecology* (Oxford: Oxford University Press, 1991); M. Paterson and M. Grubb, "The International Politics of Climate Change," *International Affairs*, 68, no. 2, 1992. For a somewhat more realistic account of the more general problem of gestural supranational environmental politics, see Michael G. Huelshoff and Thomas Pfeiffer, "Environmental Policy in the EC: Neo-Functionalist Sovereignty Transfer or Neo-Realist Gate-Keeping?," *International Journal*, 47, Winter 1991–1992.

15. There has been a recent trend within state theory and comparative political economy to deploy the distinction "strong state," "weak state" to inform an analysis of the different strategic potentialities of individual states. See inter alia, James G. Midgal, *Strong Societies and Weak States* (Princeton, N.J.: Princeton University Press, 1988); Bertrand Badie and Pierre Birnbaum, *The Sociology of the State* (Chicago: University of Chicago Press, 1983); and Michael M. Atkinson and William D. Coleman, "Strong States and Weak States: Sectoral Policy Networks in Advanced Capitalist Economies," *British Journal of Political Science*, 19, 1989.

16. A strategy perhaps most clearly pursued by the both Japan and the United States in Southeast Asia. See K. Doershug and A. Schuff, "Opportunity from Waste: The Case of Taiwan," *Journal of Southeast Asia Business*, 7, no. 2, 1991; and J. Quackenbush and K. Rushton, "The Environmental Protection Industry in ASEAN," *Journal of Southeast Asia Business*, 7, no. 2, 1991.

17. R. J. Samuel, *The Business of the Japanese State: Energy Markets in Comparative and Historical Perspective* (Ithaca, N.Y.: Cornell University Press, 1987).

18. J. Bongaerts and D. Heinrichs, "Government Support of Clean Technology Research in West Germany: Some Evidence," *R & D Management*, *17*, no. 1, January 1987; H. P. Lorenzen, *Effektive Forschungs—und Technologiepolitik—Abschäetzung und Reformvorschläege* (Frankfurt: Campus, 1985).

19. Joachim Hirsch and Roland Roth, *Das Neue Gesicht des Kapitalismus* (Hamburg: VSA, 1985).

20. On this global economic dynamic, see Michael Aglietta, "Capitalism in the Eighties," *New Left Review*, *132*, 1982; and also see Martin O'Connor, "On the Misadventures of Capitalist Nature," Chapter 7, this volume.

21. It is important to note here that the state is not a necessarily unified homogeneous agent, and that different institutions within the state apparatus are likely to employ somewhat different modes of risk assessment, reflecting the specificity of the way in which particular risks impinge upon particular institutions. Nonetheless, a general mode of risk assessment can be identified that is the outcome of the various struggles within the state over the relative cost and consequence of specific risks.

22. Anthony Giddens, *The Consequences of Modernity* (Cambridge: Polity, 1990), pp. 124ff.

23. The global warming potential of CFCs is between 1,500 and 7,300 times that of carbon dioxide on an equivalent weight basis. Thus, it has been estimated that CFCs account for around 10% of the warming impact over the previous 100 years; see Scott Barrett, "Global Warming: Economics of a Carbon Tax," in David Pearce, ed., *Blueprint 2: Greening the World Economy* (London: Earthscan, 1991).

24. Note, however, the carcinogenic properties of benzene, the most keenly advocated alternative propellent.

12

Discourse and Power in Environmental Politics: Public Hearings on a Bleached Kraft Pulp Mill in Alberta, Canada

Michael Gismondi
Mary Richardson

> Most calculations of risk disregard uncertainties, particularly for populations at special risk. The very young, the very old, and the health-impaired are seldom taken into account in calculations of risk. The important question is: Whose health are we not protecting?
>
> Reactive behavior in relation to toxic chemicals is considered to be immoral because it impinges on the rights of humans; on the rights of future generations; and on the rights of other species to freedom from environmental insult.
>
> The use or nonuse of toxic chemicals is not a question that science can resolve. Because it is primarily a moral question, its resolution will not wait for all the facts to come in.[1]

1. Introduction

This chapter is part of a larger study whose purpose is to examine the efficacy of the environmental public hearing process in Canada for allowing the direct and free expression of environmental values

Professor of Sociology and Professor of Philosophy, respectively, at Athabasca University, Alberta, Canada. An earlier version of this chapter was presented to the Canadian Studies Association symposium on Ecology and Culture in Canada, in Victoria, B.C., May 1990. First published in *CNS, 2(3)*, no. 8, October 1991, pp. 43–66.

by individuals, communities, and citizens' groups. A recent Government of Canada report, *Public Review: Neither Judicial, nor Political, but an Essential Forum for the Future of the Environment*, champions the public hearing process as "a service requested of the public by the government to help it make an informed decision and to favor a harmonious relationship between economic and environmental protection." For the authors of this report the public hearing serves in a transparent way as a means of defining "the values which the population associates with a specific proposal" and "a forum in which expert opinions on technical subjects as well as value judgments or the choices of society may intersect and merge."[2] We find, however, that the extent to which the public hearing performs such functions is open to question. Carol Cohn argues that professional discourses can define "how issues are thought about," "who may be heard as legitimate," "what may be credibly said," and "what must remain unspoken."[3] In her view, the outcome is control over public debate and the harnessing of popular discontents. In our analysis of the transcripts of an environmental public hearing, we note that the public appears constrained by, or struggling within, the ways of speaking and thinking that experts and project proponents have developed to debate the environmental impact of their projects. But we have also discovered a variety of instances in which hearing participants spoke in their own voices and successfully questioned the social authority of the dominant discourse while resisting its narrow assumptions and restrictive conventions.

This chapter counterpoints the opposing discourses of representatives of Alberta–Pacific Forest Industries (ALPAC) and opponents of their proposed bleached kraft pulp mill. (Bleached kraft pulp mills typically use chlorine or chlorine dioxide and caustics to bleach or whiten their pulp. Mills that bleach with chlorine are responsible for releasing a large number of chlorinated organic compounds, including toxic dioxins and furans, which are persistent in the environment and biomagnify up the food chain.) Our account highlights "transitional moments" when interventions by environmentalists demystified and unmasked established norms and understandings, emphasizing how the interests that underlay the language conventions used by ALPAC and their expert witnesses were exposed, and how the course of the hearings was altered. The transitions we identify mark instances when the ALPAC Review Board clearly took notice (as reflected in its report) of certain alternative understandings and arguments that called into question previous ways of speaking and thinking.

In our judgment, there was a development of understanding and sophistication of argument throughout the hearings that alerted the

ALPAC Review Board members and the public to the use of language by the proponent, making possible collective criticism of the environmental, scientific, and economic premises of the ALPAC project.

2. Economic and Political Background

In December 1988 the premier of Alberta announced that ALPAC had been given "approval in principle" to build the world's largest single-line bleached kraft pulp mill along the Athabasca River in the small farming community of Prosperity in Athabasca County, Alberta. At a cost of $1.3 billion, the Japanese-owned mill would produce 1,500 air-dried tons of pulp per day from logs drawn from a Forest Management Area (FMA) of 73,000 square kilometers of boreal mixed-wood forest (or 12% of the Province of Alberta, an area approximately the size of Austria).[4]

The ALPAC proposal brings to seven the number of pulp mills on the Peace–Athabasca watershed: two older kraft mills at Hinton and Grande Prairie, one kraft mill recently constructed at Peace River, two chemi–thermo–mechanical (CTMP) mills at Whitecourt, another at Slave Lake, and the proposed ALPAC mill in Athabasca.

At the time of the hearings, it was expected that ALPAC alone would add some 1,500 kilograms per day of organochlorides (AOX) to the Athabasca River, bringing to 5 tons per day the amount of organochlorides flowing into the feed waters of the Peace–Athabasca delta.

The provincial government conducted a partial environmental impact assessment (EIA) of the ALPAC proposal in the spring and summer of 1989. When ALPAC was unable to address many questions raised by environmentalists and civil servants, rising public pressure and the threat of litigation compelled the federal and provincial governments to jointly constitute the ALPAC Review Board. The Province of Alberta was dragged kicking and screaming into this escalating environmental review, the most comprehensive scrutiny of a pulp mill ever conducted in Canada.

The ALPAC Review Board had numerous constituencies represented in its membership: four provincial government appointees (a local farmer, a local airline owner, the local school superintendent, and the chief of the Fort McKay Indian Band); two federally appointed scientists (an internationally respected water quality scientist, who is Killam Scholar, and a professor of environmental science with expertise in EIA procedures); and a representative of the government of the Northwest Territories (NWT). The chairperson was the head of the Energy Resources Conservation Board of Alberta.

Significantly, public pressure also led to the board's purview being expanded to include examining "the cumulative effects on the Peace–Athabasca River system of . . . existing discharges as well as those which would result from the ALPAC and other proposed mills." The board's purview, however, did not include the environmental impact of timber harvesting in the proposed FMA area.[5] The hearings took place in 11 communities in northern Alberta and the NWT in late 1989.

In March 1990 the ALPAC Review Board recommended in its report that the mill "not be approved at this time." While recognizing that the ALPAC mill was "one of the least polluting bleached kraft pulp mills in the world," and noting that "it would have economic benefits associated with it," the ALPAC Review Board recommended that "further scientific studies on the river systems be conducted to determine if the Alberta–Pacific proposed mill could proceed without serious hazard to life in the river and for downstream users."[6] Review Board members estimated 3 to 5 years for the impact studies to be carried out properly.

3. Discourse and Counterdiscourse

Roger Anderson believes that "behind many of the variations in language use lie differences in social power," and argues that "we must concentrate more on the variations in language and how they relate to actual users in specific situations."[7] For him, "social power can be created and justified through discourse and practices shaped by language."[8] Bruce Lincoln, distinguishing between force and discourse in *Discourse and the Construction of Society*, explains the power of discourses in this way:

> Discourse supplements force in several important ways, among the most important of which is ideological persuasion. In the hands of elites and of those professionals who serve them (either in mediated fashion or directly), discourse of all forms—not only verbal, but also the symbolic discourses of spectacle, gesture, costume, edifice, icon, musical performance, and the like—may be strategically employed to mystify the inevitable inequalities of any social order and to win consent of those over whom power is exercised, thereby obviating the need for the direct coercive use of force and transforming simple power into "legitimate authority."[9]

Foucault presents these questions in a different way when he explains that discourses are not repressive but "permissive" ways of exercising power over us as individuals. According to Foucault, "The exercise of power consists in guiding the possibility of conduct and putting in order the possible outcome. . . . Basically power is less a confrontation between two adversaries or the linking of one to the other than a question of government. . . . To govern, in this sense, is to structure the possible field of action of others."[10] Discourses exhibit two combinations of governing effects: (1) totalizing techniques that privilege the knowledge, competence, and qualifications of one social group; and (2) individualizing techniques of power or "a form of power that makes individuals into subjects."[11]

Drawing upon an analysis of the confessional aspects of power that emerged from Christianity, Foucault calls the latter type of power "pastoral" power: "This form of power cannot be exercised without knowing the inside of people's minds, without exploring their souls, without making them reveal their innermost secrets. It implies a knowledge of the conscience and an ability to control it." For Foucault, the "modern State" is a new organization of this kind of pastoral power that integrates the individual into the state "under one condition, that this individuality would be shaped in a new form and submitted to a set of very specific factors."[12] Ashforth uses Foucault to critique commissions of inquiry, such as the ALPAC Review Board, as mere regimens of power.[13] He believes that public hearings extend the knowledge/power of the state into the innermost consciousness and secret soul of social activists (environmentalists?) in order to control them. We, like many respectful critics of Foucault,[14] are not persuaded that this submission of consciousness and the soul to the state's regimes is necessarily complete. In particular, we disagree with Ashforth, and feel that new forms of subjectivity arose in the process of declaring difference and refusing integration and domination during the ALPAC hearings.

For that reason, when examining the ALPAC hearing transcripts we are concerned neither solely with the rhetorical power of language, nor simply with its power to distort reality or to hide the proponent's efforts to present reality in a particular way. On the contrary, this chapter is about moments in the ALPAC hearings when less powerful groups undermined the discourses empowered by the dominant groups and in the process constructed counterdiscourses of their own. However, we do not read these social or ideological confrontations in the same way we would read texts. These are not confrontations lost or won simply on the battlefield of meanings and values. The acts by subordinate groups, of questioning convention, subverting dominant discourses, and asserting counterdiscourses, are highly political. They occur in the midst of complex sociological and historical processes.[15]

Thus, the joint provincial–federal ALPAC review hearing process itself must be understood within the particular institutional and social context of Canada in the late 1980s (where provincial–federal power struggles over environmental jurisdiction were fundamentally altered when the Federal Court of Canada forced a federal review of the Province of Saskatchewan's megaproject, the Rafferty–Alameda Dam in the middle of its construction). In 1989, in Alberta, these tensions between federal and provincial states were exacerbated by rising, organized pressure from environmental groups and opposition parties upon governments to protect and conserve the environment.

In our judgment, there was a reciprocal flow between the historical context and the hearing process which created possibilities for disruptive discourses to be heard, for new subjectivities to arise, and for a calling forth or mobilizing of organized resistance to the ALPAC proposal (e.g., Athabascans often remarked that they became environmentalists only after they heard the concerns raised when the ALPAC Review Board came to town). The ALPAC Review Board dynamic undermined the power of the state and capital (in this particular instance) by providing opportunities for differing values to be expressed, for public challenge of company claims to be exploited, and for presenting focused counterarguments that could disrupt the previously persuasive discourse of the ALPAC proposal.

We will focus here on five examples of how the public and some board members countered ALPAC's use of specific discourses intended to mystify the public (and the board) and to win consent for their kraft pulp mill proposal as a legitimate, scientifically sound, and environmentally acceptable project. The examples come from transcripts of the ALPAC hearings.[16] We choose to draw out how detractors of the proposal criticize, question, and sometimes subvert ALPAC's conventional or "totalizing" representations of the marketplace; the objective nature of science; the scientific status of "impact science"; the "universal" nature of scientific standards; and the attitudes of organized paperworkers toward the jobs–pollution trade-off.

4. The Market Made Me Do It

ALPAC maintained throughout that it was responding to impersonal market forces. Because its clients wanted bright white paper with strong fibers, ALPAC was forced to bleach their pulp with chlorine.

Likewise, market requirements for strong white fibers explained why, though ALPAC was supportive of recycling in principle, it could not mix in recycled fibers, because this would lower the whiteness of pulp made from Alberta's virgin hardwood fiber.

> *Ms. Peruniak (Housewife)*: What do you base your market research on?
> *Mr. Fenner (ALPAC)*: . . . The market itself.
> *Ms. Peruniak*: . . . I put it to you that within the next decade, you are going to see major changes in the bleached kraft pulp market. And I think my estimations are probably as good as yours.
> *Dr. Schindler (Board Member)*: Mr. Fenner, to what extent does the pulp and paper industry determine these markets or try to drive them? It is my experience from soap and detergent companies and the power industry that these markets can be driven rather than predicted. They push the markets in the directions they want them to go. I once overheard one, admittedly drunken, detergent executive brag to one of his colleagues that they could package horseshit in a yellow and orange box and housewives would still buy it.
> *Ms. Peruniak*: Not this housewife.[17]

ALPAC's Mr. Fenner does not define the complex word "market." But the word "market" empowers his discourse. He presents ALPAC's case as if the company made no choices or had no choice when it made a choice: no choice targeting its market niche; no choice of bleaching process (i.e., chlorine bleaching over other processes less damaging to the environment); no choice but to reject recycling. The discourse of the market is persuasive because ALPAC wants to appear subordinated to market laws and market rules carved in stone.

The "impersonal," "objective" appearance of market forces combines with the already abstract corporate persona of ALPAC to further remove from public awareness the agency implicit in company decision making. Because it did not create the market, the assumption is that ALPAC simply responds helplessly to it. Following Mr. Fenner's logic to the end, it is consumers who actually forced ALPAC to pollute because "the market" is you and I, the consumers of shiny white paper. We are the cause of pollution.

In the "theater of the hearing process," Schindler speaks out of character and breaks through this dominant discourse. His use of humor and anecdote pierces through the "authority" of Mr. Fenner's representation of ALPAC as a victim of market forces. He reestablishes the truth that corporate decision makers are not victims of an impersonal marketplace. His intervention reveals agents with ALPAC who make market choices, choices about what they produce and how. The choice to use chlorine bleaching to make white paper is recognizable now as a grab for a highly profitable end of the paper market. The choice not to incorporate recycled paper also means ALPAC has made a choice to pollute in order to make money. Schindler announces to ALPAC that he will not accept "economic" rationales as a justification for evading responsibility for environmental damage.

But Schindler's anecdote does more than deprive ALPAC decision makers of the mask of "the myth of the market." Schindler advances the same counternarrative *voiced by ordinary Canadians* critical of ALPAC's proposal: that corporations could, if they chose to focus their efforts in this direction, create a market for unbleached or less white paper with recycled fiber content (and thus eliminate chlorine and its significant environmental damage). By unmasking the myth of an impersonal market, he reasserts the truths that corporate salesmen create markets, and that they can and do manipulate consumers.

Finally, Ms. Peruniak's retort of "Not this housewife," directed at both Fenner and Schindler, asserts the sovereignty of the thinking consumer. She announces the presence in the 1990s of environmentally conscious persons who refuse to be victims of the market or of corporate decision makers.

5. The LC50 Test: The Man Who Loved Fish

The LC50 is a lab test for measuring the acute lethal toxicity of industrial effluents to fish or other aquatic organisms. According to a government report, "Samples of effluent identified as 'non-toxic' are those in which more than half (50 percent) of the test fish exposed to full-strength effluent for 96 hours survived."[18] That is, if ALPAC's undiluted pulp mill effluent kills slightly less than 50% of the test fish within 96 hours, its effluent passes the test.

To this point in the review of the project, the public had accepted without question a "technologist" understanding of scientific tests. The LC50 appeared in the EIA as simply a set of procedures agreed upon and set out in a governmental policy manual. The government technologist who reviewed the ALPAC data checked to see that the company had followed the required procedures. The scientist's job was to report the results of the test "on the stand" to the panel. Either the effluent passes or fails the test. The theatricality of the scientists' expert testimony was much like that of a coroner at a murder trial. ALPAC's data suggested that it would routinely pass this test, unlike many British Columbia pulp mills. Alberta Environment agreed. But Dr. Ian Birtwell (of the Department of Fisheries and Oceans Canada), drawing from his research on fish living downstream from kraft pulp mills on the Fraser River in British Columbia, alters the discourse.

Dr. Birtwell: My main area of research is on the sub-lethal effects of contaminants, that is, levels that don't inherently kill fish outright, but perhaps may have an effect at a level below that, that would have killed them. . . . Many times an animal may be debilitated and essentially ecologically dead by conditions far below those which kill it in four days, and kill 50 percent of them in four days. Surely what we are trying to do is protect that level of population, the zero to 50 percent that is sensitive. In regulatory bioassays where effluents are exposed to fish, typically one may consider the effluent as being acceptable if it does not kill 50 percent of the test fish within four days. Well, from my perspective, it's those zero to fifty percent which are equally important. For biological purposes, we've typically used the 50 percent response level, but I think we also have to consider response at the 1 percent, the 5 percent and the 10 percent level. Those fish, or whatever animal we are studying and testing, is also of relevance in the environmental context.[19]

Here Birtwell questions the test itself as a measure of nontoxicity to fish. He calls into question not the procedures, but the underlying assumptions of the test. In fact, he suggests that while ALPAC's effluent passes the LC50 test, "passing" does not indicate that ALPAC's effluent is harmless to fish.

But his discourse does much more. His "love of fish" reveals a subjective aspect of science that was stunning after months of dry, passive discussions of whether the mill's effluent would pass the LC50 and cause no harm. This discourse breaks us free from two conventions in science writing: the present tense and the passive voice. For Andersen, the use of the present tense supports the idea of science as the objective discovery of timeless truths. It does not let the listener understand that the scientist has a part to play in "creating and re-creating" science.[20] For Killingsworth and Steffens, the use of the passive voice in science writing "obliterates agents of actions and thereby obscures responsibility and/or authority."[21]

Birtwell breaks us free from this timeless, object–object way of seeing science as value-free with no actors creating and re-creating it. Instead of reducing the role of the observer and his or her interaction with the fish and the text, Birtwell's subject–object intervention enlarges human agency and responsibility. On the one hand, the word "we" removes the distance created by a scientific discourse that devalues human emotion and ethical principles; on the other hand, it affirms that even in a value-free scientific approach, values are expressed.

The emphasis in the ALPAC EIA was that scientific facts spoke for themselves and expressed facts alone, not values. In the event that values were expressed, they were scientific conventions and somehow above political and ethical criticisms, that is, a neutral discourse, not the decisions of ALPAC's "hired" scientists. In contrast, Dr. Birtwell's expression of emotion was significant. ALPAC public relations people often suggested that public opposition to their project was based on emotion and a "fanatical," "tree hugger" love of the environment. ALPAC "spin masters" often labeled criticisms of their project as emotional in order to devalue them. This technique can be called "derogatory semantics"; discourses that are marked emotional are marked negatively. In the hearings, scientific interventions that are marked emotional are marked negatively in a specific way, that is, as nonscientific.

To this point in the struggle over the discursive terrain, whenever ALPAC showed a human face it enhanced its positions. ALPAC spokesman Gerry Fenner used emotion to his advantage when, after giving hours of sober responses based on the "facts," Fenner told parents concerned for their children's health that he was a parent too, that he fished with his kids downstream from the pulp mill he managed in Skookumchuck, B.C., and that his family enjoyed eating the fish they caught.[22] In contrast, when local people showed emotion, their points were devalued as nonrational appraisals of the project. Dr. Birtwell, a representative of the Government of Canada and a scientist and specialist studying the effects of pulp mill effluent on fish, made his emotional expression of the love of fish and changed all that.

6. Natural Science versus Impact Science

Dr. Bill Fuller, a member of the Friends of the Athabasca Environmental Association (FOTA), chose to discuss the quality of the science in the ALPAC EIA. Dr. Fuller is Professor Emeritus and former Chair of the Department of Zoology at the University of Alberta. He has participated on federal environmental assessment review panels, worked for years in the Peace–Athabasca delta, and chose to retire outside Athabasca "to be near a northern river."

Dr. Fuller prefaced his presentation with a rhetorical question:

> What in all the mass of material submitted by ALPAC . . . what in all the hundreds of pages that I read would be accepted for publication in a peer review journal if it were submitted to me? The only answer I can come up with is nothing, zero, zilch, dick all, however you want to express it.[23]

Now these feelings could easily be dismissed as the words of a crusty scientist who was bitter about losing his retirement paradise. Bill Fuller was aware of that, and chose then to read from an editorial that had appeared in *Science* (May 7, 1976). The editorial was entitled "Impact Statement Boondoggle" (at this point, Dr. David Schindler recognized the title and began to laugh openly):

> Many politicians have been quick to grasp that the quickest way to silence critical "ecofreaks" is to allocate a small portion of funds to any engineering project for ecological studies. Someone is inevitably available to receive these funds, conduct the studies, regardless of how quickly results are demanded, write large reports containing reams of uninterpreted and incomplete descriptive data, and in some cases, construct "predictive" models irrespective of the quality of the data base. These reports have formed a gray literature of reports . . . voluminous and so limited in distribution that their conclusions and recommendations are never scrutinized by the scientific community at large. Often the author's only scientific credentials are the impressive title in a government agency, university or consulting firm. This title, the mass of the report, the author's salary, and his dress and bearing carry more weight with the commission or study board, to whom the statement is presented, then either his scientific competence or the validity of his scientific investigation. Indeed, many agencies have found it in their best interests to employ a "traveling circus" of "scientists" with credentials matching those requirements. As a result, impact statements seldom receive the hard scrutiny that follows the publication of scientific findings in a reputable scientific journal.[24]

Here the "situation of the discourse" works on the changes in meaning between "natural science," on the one hand, and "impact science," on the other. We have Fuller the scientist, sitting in front of two scientists on a review panel, facing five or six ALPAC scientists across the room (the traveling circus), entering this devastating critique of "impact science" into the record. The scientist who wrote this editorial is the scientist ALPAC fears the most. When Dr. Fuller reveals that Review Board member Dr. David Schindler was the author of this critique, the "situation" empowers Fuller's effort to separate the two sciences.

An important transition occurred here. Throughout the year of the ALPAC open houses and public presentations, mill boosters often abdicated the review of the EIA science either to the proponent (we trust ALPAC) or to the government (we trust our civil servants). Scientists belonging to environmental groups were less inclined to abdicate this responsibility.[25] Also, as laypeople developed sophistication, they too began to criticize the science or social science in the proposal, often asking for third-party independent verification of company studies or asking for studies that were not done because ALPAC felt they were unnecessary. By separating science from "impact science," Fuller questioned the legitimacy of impact science and privileged the doubts of even nonscientist citizens.

Finally, within the theater of the hearing process, Fuller prepared his dramatic conclusion. After arguing that natural science in general was more rigorous than the impact science in ALPAC's EIA, Fuller resolved that *real impact science* should be even more rigorous than natural science. Quoting a response to Schindler's original editorial, Fuller concluded that "science is a self-correcting endeavor . . . and one is confident that correct results will come eventually, leaving only a relatively harmless pile of worthless papers, wasted man-hours, and broken test tubes behind. But *we cannot afford to let impact science follow tradition.* The legacy will not be broken test tubes but hopelessly and permanently crippled ecosystems."[26]

7. Special Populations at Risk and the Ethnocentrism of Scientific Standards

> Alberta–Pacific's EIA says dioxions and other chlorinated organics accumulate in the fatty tissues of fish and fish livers. The EIA goes on to say that human ingestion of chlorinated organics would be slight because people don't eat fish liver. Well, people in our region do eat fish liver. In fact, loche on burbot liver is considered a delicacy.[27]

The debate on the downstream effects of pulp mill effluents often appeared as if it were simply a scientific argument over the presence or absence of dioxins and furans in ALPAC's pulp mill effluent. Two of the most toxic pollutants were the focus of concern: 2,3,7,8-TCDD dioxin and 2,3,7,8-TCDF furan. These are produced in significant quantities in older bleached kraft mills. Reputedly, 2,3,7,8-TCDD is the most highly toxic substance ever produced.[28]

Much debate centered on the persistence in the environment of these highly toxic dioxins and furans found in the effluent ALPAC would put in the Athabasca River.[29] Dioxins and furans have long half-lives in the ecosystem. They bioaccumulate and biomagnify in fish, especially in fish liver. How they may affect mammals up the food chain, including humans who eat fish, is hotly disputed. ALPAC brought toxicologists to testify who argued for acceptable thresholds (or amounts of dioxin) that could safely be ingested or taken up into human systems.[30]

But the government of the NWT made a presentation using highly respected scientists who changed the frame of reference of this "dioxin threshold" discourse. They raised the issue of a "special population at risk," the native people of the NWT whose main source of protein is "country food." The board was told that "Health and Welfare Canada's permissible concentrations of 2,3,7,8-TCDD in fish for human consumption are based on *assumed rates of fish consumption* that are far lower than those of Aboriginal Peoples in the lower Athabasca, Slave and Mackenzie River systems." For example, the weekly consumption of fish by "many Aboriginal Peoples [living along the Slave River which will be downstream from four Alberta kraft pulp mills by 1993] may be tenfold higher than that of average Canadians."[31] Furthermore, natives in the NWT[32] usually eat the whole fish and *particularly enjoy fish liver*.

Dr. Swain, the NWT's expert on toxicity, presented evidence on the presence of PCBs (a related molecule whose equivalent toxicity is 40% that of dioxin) in the systems of families of Great Lakes fishermen who ate fish fillets 1 to 3.5 times a week. Swain's studies of these families, especially the pregnant mothers and newborn children, showed that "we are dealing with a unique set of toxicants . . . affecting not only the individual exposed, but the progeny that the individual bears." He found negative health effects such as lower IQ, poorer fine motor skills, lower sperm counts, and more. Swain concluded, "What I can tell you is that very critical human tissues are being affected by toxic substances. And this is not a special population at risk."[33]

> *Dr. Swain (NWT Scientist)*: The majority of standards that have been developed . . . are based upon the average consumption pattern of the average human being . . . consuming something like between one and half and three and half grams of fish a day, or . . . something in the order of one fish meal per week for recreationally caught fish.
> *Dr. Ross (Board)*: Perhaps I could just make sure I understand how . . . I might apply correction factors for these special populations. My assumption is that if a population eats, for example, 25 times more fish than the average New York resident, then I ought to reduce any standard by a factor of 25 to compensate for that; is that correct?
> *Dr. Hallett (NWT Scientist)*: That's correct.
> *Dr. Swain*: . . . In fact, 25 is probably an extremely low number.
> *Ms. Gilday (Board Member from NWT)*: . . . The kind of studies that you are talking about are very ethnocentric as far as standards are concerned. We were in Fort Resolution where we were told the

man eats fish five times a week. That's a father. And the mother also eats five fish a week. That's ten fish, one week, in one family's life. Considering what you told us this morning, we are not only talking about 10 parts of dioxin, but taking into consideration PCB and all the other compounds, when you look at its long term or cumulative impact, what would you say to the people in the Northwest Territories? What do they have to look forward to?[34]

Cindy Gilday,[34] the NWT representative, was the first to utter the word "ethnocentrism," which means to social scientists the tendency to privilege (often unconsciously) the point of view of one's own cultural norms, values, and practices over those of another culture or society (which is usually judged to be inferior to one's own culture).

At issue is the "universality" of the discourse empowered by natural science. Natural science as a discipline is considered the last bastion of Western thought. Scholars have questioned the fundamental assumptions and methods of political science, history, sociology, literary studies, and many other fields, and have shown many of them to be ethnocentric or "orientalist." But the methods and principles of natural science appear to be above this criticism. They appear applicable across cultures without distortion. Their claims are based on the immutable physical properties of nature, not the mutable subjective classifications created by cultures or the nonsciences.[36]

"Scientific standards" claim the power of the discourse of "pure science" with its "neutral" qualities of experimentation and universal principles. By treating the "scientific standards on dioxin uptake" as something other than pure science, Dr. Swain and Ms. Gilday break the power of this crucial link. The fracture reveals: (1) white, Western (male?), decision makers (many of them scientists) drawing up standards for allowable levels of dioxin in fish; and (2) a place for politics because it reveals the possibility of making the choice not to add to dioxin loadings in the river.

The notion of a "special population at risk" links opposition to the mill to a powerful counternarrative available in Canada: the debate over native rights.[37] Once the ethnocentrism of dioxin standards is established, Dr. Swain's evidence concerning the intergenerational impact of toxins (i.e., behavior disorder, decreasing IQ by four points every generation; decreasing sperm counts) links the issue to a more profound and powerful discourse which speaks to the possibility of the slow, invisible genocide of native people. His conclusions put heavy pressure on the ALPAC Review Board to decide against the mill. They would quote Dr. Swain at length in their final document as a justification for their decision against the mill.[38]

8. Unions and Chlorine Bleaching

Northern Alberta is an underdeveloped area of western Canada. Community populations are in decline. Unemployment ranges from 4 to 10% higher than the provincial average and in many northern native communities unemployment is 50% or higher. In the Athabasca region, in particular, there is little industrial development and youths migrate to the cities for work. The ALPAC proposal promised 1,100 jobs throughout the north: 440 in the mill and 660 in the bush. In Canada, jobs in the forest sector are well paid because this industry is highly unionized. Spin-off jobs were expected to double the employment impact in the mill region. The government of Alberta argued that these direct and indirect jobs would diversify the Alberta economy away from its heavy reliance on farming and oil and gas extraction. Mill boosters wanted jobs so young people could stay "home" and work in their own communities, and they were ready to sacrifice some "use" of the environment for jobs. In sharp contrast, both the Canadian Paperworkers Union (CPU) and the Pulp, Paper, and Woodworkers of Canada (PPW) argued against the ALPAC proposal on environmental grounds. Both the CPU and PPW were less inclined to sacrifice the "health" of the environment or the youth of northern Alberta.

The CPU is a national industrial union of some 72,000 members. Dave Coles, the Alberta organizer of the CPU, and Keith Neuman, the national research officer, read a prepared text from the CPU national leadership that questioned the appropriateness of chlorine bleaching in ALPAC's proposed mill. The brief emphasized that they spoke on the behalf of people who have firsthand work experience inside pulp and paper bleach plants. The CPU felt that health problems among their members and evidence of environmental pollution downstream from mills indicated chlorine bleaching should be phased out in existing plants and that no new plants using this process should be built.

Dr. Schindler: So is this not tantamount to endorsing that the occupation that your members hold be rendered extinct?

Mr. Coles (CPU): Well, I guess we have to face the realities of life. I'll use for an example when people worked in asbestos mines and were assured it was good for their health, they ended up with problems and had to make changes. We as a union, and as a society, are having to wrestle with a lot of these environmental issues brought to fruition in the last years. There are problems with pulp and paper, and we appreciate that.[39]

Mr. Fenner: . . . Are you going to not try to organize this mill because of your concern about the environment?

Mr. Coles: . . . We'd never disassociate ourselves from any class of worker, whether it's in this pulp mill or any other.

Mr. Neuman (CPU): We believe that workers should be represented by unions everywhere, no matter whether we think the project that they are working on is ideal or not. So that's kind of an irrelevant question.[40]

The CPU's intervention arrested two broad public discourses often heard since the mill was announced. The first was that industrial workers were antienvironmentalists because environmentalists threatened their jobs. The CPU spoke at the invitation of FOTA and their intervention indicated that pulp mill workers are willing to work with environmentalists to promote environmentally sound development.

The CPU speakers also undermined the powerful argument that the proposed jobs would solve the employment problems of northern Albertans. Throughout, the board heard submissions from the Chamber of Commerce, the local councils, the local member of the legislature, some native association executives, and many members of the public who "knew little or nothing of the environmental issues" but trusted ALPAC and could confirm "that the region badly needed the jobs and investment." This put special pressure upon the local board members to approve the mill. If they voted against it, they would have to live with the displeasure of many of their neighbors who would accuse them of refusing an opportunity to remedy unemployment. By linking the jobs directly to a potential health hazard for mill workers, the CPU added a sobering dimension to these potential jobs: a potential health hazard for the youth of Athabasca and northern Alberta.

This report from the shop floor further sensitized the hearing board and the general public to increasing concerns about chlorine bleaching of pulp. Furthermore, because the unions appeared to be putting the interests of workers ahead of their corporate interest to recruit members and preserve the kraft pulping profession, ALPAC's corporate self-interest on the chlorine question appeared in stark relief.

Finally, Mr. Fenner's query as to whether the CPU would refrain from organizing the ALPAC plant based on environmental grounds confirmed in an ironic way how labor–capital struggles play themselves out even within this discursive field of the environmental hearing. The responsible comebacks of the CPU speakers lay bare some fundamental oppositions of labor and capital within the debate on the environment in the 1990s.

9. The Struggle over Language

Everyone—politicians, businessmen, consumers, and social activists—was an environmentalist in the 1990s. In the ideological and social conflicts concerning the environment, we find that the "self-sufficient" environmental worldview of one group is not simply replacing the nonenvironmentalist world views of another.[41] Rather, the discourses on the environment more resemble a contested terrain where meanings are fought over, shifted, abandoned, transformed (witness the abandoning of "sustainable development" by environmentalists who argue that business leaders coopted this "green" buzz word to legitimate an economic growth paradigm). As Stuart Hall explains, "Differently oriented accents intersect in every ideological sign. . . . As different currents constantly struggle within the same ideological field, what must be studied is the way they contest, often around the same idea or concept."[42]

The ALPAC hearings revealed many power splits and fractures in communities, native bands, the civil service, the state(s), and the board. Our reading only begins to suggest some ways in which these splits either constrained the free flow of language or allowed for subversion of the dominant discourse. The joint federal–provincial review process we described resembles the ideal type of commission of inquiry that Ashforth called a "reckoning scheme of legitimation." He claims that "analysis of dimensions of commission work can reveal some of the discursive formations underlying state power: ways of speaking about social life which make possible the work of organizing political subjection."[43] Yet, in our judgment, the panel process and its "constantly shifting ground" put ALPAC on the defensive. The ALPAC review was not the planned reckoning scheme of Ashforth, but instead a social project that grew in bits and pieces as political pressure for a public review, and counterevidence during that review, mounted.

From our experience in the hearings, we find Foucault's power/ knowledge schema only partially helpful because it renders analysis of "commissions of inquiry" ahistorical. Generally, it assumes that well-organized citizen "subjects" (read social activists, environmentalists, and native bands) cannot exert pressure between the fissures in the state(s) by pitting states and their internal departments against one another; by questioning the terms of reference and composition of commissions; even by subverting a commission's ability to "organize political subjection." Finally, it underestimates popular politics, resistance, and social agency,[44] as we suggest they play themselves out in the struggle over language.[45] As Mikhail Bakhtin has explained, language is already lived in and presupposes interlocution and social context; the word in language is half someone else's.[46] Early on in the ALPAC process, the word was all ALPAC's, with the full backing of the Alberta government. Months of grassroots struggles have begun to win the words back.[47]

10. Postscript

While the findings of the original review board represented a significant victory for environmentalists, the provincial cabinet and the company were still determined to go ahead with the project. The Province of Alberta responded to the Review Board report by commissioning "a review of the review" by a Finnish pulp and paper research consultant. Its findings were consistent with those of the original review board. In June 1990 ALPAC presented to the government a mitigative proposal eliminating the use of molecular chlorine in the bleaching process. (Using chlorine dioxide instead of molecular chlorine in the bleaching process greatly reduces the production of dioxins and furans.) To study the feasibility of the mitigative process, the Alberta government commissioned a scientific review of the modified process. In October 1990 the panel concluded that the process was feasible, and on December 21, 1990, the Alberta cabinet approved the ALPAC mill. The mill began operating in September 1993.

Notes

1. Dr. Jack Vallentyne for the Athabasca Tribal Council, *The Alberta–Pacific Environment Impact Assessment Review Board Public Hearings Proceedings*, pp. 4916–4917.

2. *Public Review: Neither Judicial, nor Political, but an Essential Forum for the Future of the Environment* (Ottawa: Supply and Services Canada, 1988), pp. 12, 1, 2.

3. Carol Cohn, "Emasculating America's Linguistic Deterrent," in A. Harris and Ynestra King, eds., *Rocking the Ship of State* (Boulder, Colo.: Westview Press, 1989), pp. 153–170.

4. The largest shareholder in Alberta Pacific is Crestbrook Forest Industries of British Columbia. This "Canadian" firm is owned 32% by Mitsubishi and 32% by Honshu Paper Company of Japan. For an overview on this and other general issues of political economy, see Barry Johnstone and Michael Gismondi, "A Forestry Boom in Alberta?," *Probe Post*, *12*, no. 1, Spring 1989, pp. 16–19; Andrew Nikiforuk and Ed Struzik, "Great Forest Sell-Off," *Globe and Mail Report on Business Magazine*, *6*, no. 5, November 1989,

pp. 56–59, 61; and Christie McLaren, "The Wholesale Sell-Off of Canada's Boreal Forest," *Equinox*, *53*, September–October 1990, pp. 43–55.

5. In Canada, forestry falls under provincial jurisdiction, whereas fisheries, migratory birds, interboundary waters, and native lands fall under federal responsibility. Alberta Minister of Forestry LeRoy Fjordbotten is a major supporter of the ALPAC mill and he fought to exclude analysis of forestry and wildlife impacts. The board did examine timber harvesting as it affected Indian Reserve Lands, and expressed numerous concerns. Correspondence between the ministers of environment acquired under the Canadian Freedom of Information Act indicates that the federal minister initially wanted forestry included, but the province objected.

6. Alberta–Pacific Environmental Impact Assessment Review Board, *The Proposed Alberta–Pacific Pulp Mill: Report of the EIA Review Board* (Edmonton, Alberta: Alberta Environment, 1990). See Executive Summary.

7. Roger Andersen, *The Power and the Word: Language, Power and Change* (London: Paladin Grafton Books, 1988), Preface.

8. Ibid., p. 285.

9. Bruce Lincoln, *Discourse and the Construction of Society: Comparative Studies of Myth, Ritual and Classification* (New York: Oxford University Press, 1989), pp. 4–5.

10. Michel Foucault, "The Subject and Power," *Critical Inquiry*, *8*, Summer 1982, pp. 789–790.

11. Ibid., p. 783.

12. Ibid.

13. Adam Ashforth, "Reckoning Schemes of Legitimation: On Commissions of Inquiry as Power/Knowledge Forms," *Journal of Historical Sociology*, *3*, no. 1, March 1990, pp. 1–22.

14. See the critique of power in Foucault's work by P. Willis and P. Corrigan, "Orders of Experience: The Differences of Working Class Cultural Forms," *Social Text*, 7, Spring–Summer 1983, pp. 85–103.

15. Some of these are the expansion of the forest industry into previously unused hardwood resource areas; high prices for premium pulp; the globalization of markets; the Canada–United States Free Trade Agreement; the Alberta government's open invitation to offshore capitalists such as Mitsubishi to develop the northern boreal forest; the personalities and political aspirations of the respective environment ministers; the paramilitary nature of the forestry department and its "general"; federal–provincial relations and the Canadian constitutional crisis. These are counterpointed by a growing global concern about environmental problems: the greenhouse effect; global deforestation; dioxin and furan contamination from pulp mills; the closing of the shellfish fisheries in British Columbia; and a rise in the number of people in organized environmental and public advocacy groups. In the midst of this is the coincidence of a university in Athabasca, the retirement of a couple of scientists in the impact area, and a well-organized New Democratic Party association in the riding where ALPAC will build.

16. *ALPAC Hearing Transcripts*, and written submissions to the board, are available at the Athabasca University Library, Box 10,000, Athabasca, Alberta T0G 2R0, Canada.

17. *ALPAC Hearing Transcripts*, pp. 3336, 3338, 3339.

18. Environment Canada, *Aquatic Toxicity of Pulp and Paper Mill Effluent: A Review* (Government of Canada, April 1987), p. 36.

19. *ALPAC Hearing Transcripts*, pp. 2799–2801.

20. Andersen, *The Power and the Word*, op. cit., p. 147.

21. M. Jimmie Killingsworth and Dean Steffens, "Effectiveness in the Environmental Impact Statement: A Study in Public Rhetoric," *Written Communication*, 6, no. 2, 1989, p. 159.

22. Government of Alberta ALPAC, EIA, *Transcripts for the town of Boyle*, June 1989.

23. *ALPAC Hearing Transcripts*, p. 2951.

24. *ALPAC Hearing Transcripts*, pp. 2951–2952.

25. Many scientists opposed to the project work with FOTA, live in Athabasca, and come from Athabasca University. This added a tense community and institutional dimension because town and county councils supported the mill.

26. *ALPAC Hearing Transcripts*, pp. 2952–2953.

27. Frank Pope for the Shihta Regional Council, *ALPAC Hearing Transcripts*, p. 5376.

28. *Report of the EIA Review Board*, op. cit., p. 22. The chemical 2,3,7,8-TCDF is considered to be tenfold less toxic than 2,3,7,8-TCDD although its concentration is usually tenfold higher in pulp mill effluent.

29. Because the proposed ALPAC process would produce effluent with "non-detectable levels of dioxins in parts per quadrillion," the definition of "nondetectable" also became a hotly debated scientific issue. Uncertainty about ALPAC's impact on the Athabasca increased because all the evidence mounted by the Department of Fisheries and Oceans came from comparative studies of mills in British Columbia (which have no secondary treatment and much higher loadings of organochlorides and detectable levels of dioxin than ALPAC's mill). Few studies of the Athabasca exist.

30. Here the evidence is very inconclusive or, for complex reasons, very muddied. For example, the U.S. Environmental Protection Agency has a much stricter position than Health and Welfare Canada. The positions demonstrate different philosophies about dioxin standards (i.e., Canada: allow pollution up to known limits of contamination vs. United States: protect against and prevent further contamination of the ecosystem). Recent reports suggest falsification of earlier evidence downplaying the links between human health problems and agent orange and/or dioxin exposure.

31. *Report of the EIA Review Board*, op. cit., pp. 22–23.

32. See E. E. Wein, *Nutrient Intakes and Use of Country Foods by Native Canadians near Wood Buffalo National Park*, Ph.D. diss., University of Guelph, Ontario, 1989. It should be noted that for toxicological concerns not the mean consumption but the extremes of consumption should be the focus.

33. *ALPAC Hearing Transcripts*, pp. 5871, 5880.

34. *ALPAC Hearing Transcripts*, pp. 5885, 5916–5917.

35. Ms. Gilday is a Slavey from Fort Franklin, NWT. Dr. Hallett discovered dioxins in 1980 in the Great Lakes. His laboratory also did the dioxin studies on the west coast of British Columbia which led the Canadian

government to close shellfish fisheries downstream from kraft mills. At present, studies downstream of the pulp mill at Hinton, on the Athabasca, show combined concentrations of dioxins and furans in fish fillets are above the 20 ppt permissible limits for human consumption of Health and Welfare Canada.

36. We have profited from the feminist critique of science by, among others, Sandra Harding.

37. See the references in Michael M'Gonigle, "Developing Sustainability: A Native/Environmentalist Prescription for Third Level Government," *BC Studies, 84*, Winter 1989–1990, pp. 65–99; Catherine Shapcott, "Environmental Impact Assessment and Resource Management, a Haida Case Study: Implications for Native People of the North," *Canadian Journal of Native Studies, 9*, no. 1, 1989, pp. 55–83.

38. *Report of the EIA Review Board*, op. cit., p. 24.

39. *ALPAC Hearing Transcripts*, p. 3261.

40. *ALPAC Hearing Transcripts*, pp. 3275–3276.

41. This way of thinking about the tension between dominant and popular uses of the same language elements comes from Stuart Hall, "Religious Ideologies and Social Movements in Jamaica," in R. Bocock and K. Thompson, ed., *Religion and Ideology: A Reader* (Manchester: Manchester University Press in association with the Open University, 1985).

42. Stuart Hall, *The Hard Road to Renewal* (London: Verso, 1988), p. 9. We find some of the conceptual tools developed by Hall helpful for extending both Marxist and non-Marxist notions of ideology.

43. Adam Ashforth, "Reckoning Schemes of Legitimation: On Commissions of Inquiry as Power/Knowledge Forms," *Journal of Historical Sociology, 3*, no. 1, March 1990, p. 17.

44. Some instances around and through the hearings were public pressure on federal and provincial governments, national media pressure, education of the public and public servants, and the threat of a legal decision against the federal government (following Rafferty–Alameda) combined with the "realpolitik" of community environmentalism by FOTA and other mill opponents which gave voice to another reading of the mill's impact on the environment. The struggles over language in the hearing were only possible because of these social and political acts (and a range of economic factors) which occurred within a particular set of historical possibilities in late 1989 and early 1990.

45. Palmer argues that most discourse analysis fails to interpret text in context. See the critique of Foucault, and discourse analysis in general, in Bryan Palmer, *Descent into Discourse: The Reification of Language and the Writing of Social History* (Philadelphia: Temple University Press, 1990).

46. See Ken Hirschkop, "Bakhtin, Discourse and Democracy," *New Left Review, 160*, November–December 1986, pp. 92–113.

47. For a fuller treatment of themes discussed in this chapter, see Mary Richardson, Joan Sherman, and Michael Gismondi, *Winning Back the Words: Confronting Experts in an Environmental Public Hearing* (Toronto: Garamond Press, 1993).

13

Ecological Crisis and the Future of Democracy

Alex Demirović

1. Introduction

The precarious relationship between today's ecological crisis and democracy demands a response from democratic theory. In this chapter two approaches to an ecologically sound constitutional theory are discussed. I argue that both prove to be insufficient, above all in the way they construct ecological models in relation to the project of building a more democratic society. I then suggest an alternative democratic theoretical response to today's ecological crisis, relying on a concept of society's socially constructed relationships to nature (*Begriff gesellschaftlicher Naturverhältnisse*). This response involves a democratization of the "logic of differentiation" that characterizes industrial societies.

The welfare state has been in crisis for a long time. This crisis is the result of internal contradictions and the dynamic of class conflict, which in the past decades assumed the form, more or less successfully, of a compromise-oriented participation on the part of the working class in the making of social decisions. The essential characteristics of this compromise included continued economic growth powered

Department of Social Sciences of the Johann Wolfgang Goethe-Universität and the Institut für Sozialforschung, both in Frankfurt, Germany. Thanks to members of the Frankfurt Institute of Social Ecology, and to Juan Martínez-Alier, John Ely, and Michael Schatzschneider for their criticisms and useful suggestions. A version of this chapter, as translated by John Ely, first appeared in *CNS*, *1(2)*, no. 2, Summer 1989, pp. 60–61. The original German version appeared as "Ökologische Krise und die Zukunft der Demokratie," *Prokla*, *84*, no. 3, September 1991, and has been revised and translated from the German for this volume by Ron Faust.

by rising disposable income and consumption; the formation of "normal life histories" with a "collective-binding" power; and welfare state guarantees of social security. This compromise took the form, at least in the Federal Republic of Germany (FRG), of corporatist agreements between the state administration, trade union and employer associations, and political parties. For the organizations representing the working class, this way of pursuing their interests led to a growing standardization of these interests, a bureaucratization of trade unions, and a transformation of the Social Democratic Party (SPD) into a broad-based, multi-interest mass party.

The crisis characteristic of a society produced by means of welfare state intervention unfolds with a different rhythm in each different sector of society. This is the case for the entire sector of "conditions of production," as has been brought out by the new social protest movements. Three main features can be distinguished.[1] First, the Keynesian–Fordist compromise rested on an expansion of demand, a broadening of the consuming classes, and a progressive growth in productivity, all of which has led in large degree to an irrational use of nonrenewable resources. Second, state investment and social programs (e.g., housing construction, highway construction, programs to boost agricultural productivity, and so on), whose aim was to realize and maintain a class compromise organized by the welfare state, have resulted in destruction of ecological systems. Third, in the case of numerous production processes and the products themselves, the ecological, social, and individual consequences have not been considered at all or only to a limited degree.

The crisis of the welfare state can be seen especially clearly in the fact that its means of intervention, its mechanisms of coordination, and its forms of participation, all of which were constructed over recent decades, are failing in the face of the new kinds of social problems produced by the very class compromise organized by the welfare state. Above all, the different interest positions of the parties to the compromise have shifted, and with them the boundaries of social conflict. For capital, the compromise of the past decades has become a fetter: the priorities of the welfare state, articulated in its subsidy, investment, and social policies, prevent capital's flexible reaction to the crisis; to the emergence of new markets; and to the introduction of new technologies. Meanwhile, the trade unions have reacted defensively, trying to protect earlier achievements, but even they are forced to recognize that the old strategies have lost their force in the face of new constellations of power.

In short, traditional forms of state intervention are in many respects no longer desired by capital because they tend to maintain old conditions. In addition, however, the legal and fiscal means of regulation employed by political and administrative policymakers can no longer keep up with the perverse effects of their own attempts at regulation. Public policy tends toward the irrational—and here little has changed since Marx first commented on the self-referential character of state bureaucracies—because it does not know "which society it is dealing with."[2]

The crisis of the welfare state, therefore, consists of at least three overlapping crises in process: a crisis in the conditions of production; a crisis in the balance of forces leading to a class compromise; and a crisis in (the means of) state intervention.

2. Democratic Participation and the Temporal Horizons of Social Planning

With the crisis of public policy, state-centered forms of democratic public discussion and collective decision making (*Willensbildung*), in which numerous, different social points of view are brought to bear on governmental and administrative decision-making processes, are also becoming precarious. This is a problem in part because legalistic means for the self-regulation of society become irrational to the extent that they are incapable of reaching numerous sectors of society (production processes; products; collective and individual consumption) where ecological problems arise. Legal norms, and the administrative interventions derived from programs based on these norms, seem, in the face of complex technologies and problems defined strictly in technical–scientific terms, to be simply antiquated.[3]

But more important than this practical dimension, although closely related to it, is the fact that capitalist society escapes temporal self-reference by means of a legally codified form of democratic public discussion and collective decision making.[4] While ecological damage, risks, and dangers have a temporal range of many thousands of years, the temporal horizon of current, more or less democratic procedures is restricted to perhaps two or three legislative periods. And since this time frame pales even in terms of the planning horizons of industry, one can hardly speak of a legal regulation, much less of a democratically determined regulation, of the social appropriation of nature (which, of course, takes place today in an almost exclusively industrial-capitalist form).

As far as the future of democracy is concerned, then, we are dealing with a problem of two temporal horizons—one political, the other ecological—that at present are on the verge of uncoupling from one

another. Only when there is a solution to the problem of the length of the
temporal horizons that frame all forms of society's socially constructed
relationships to nature will democracy—that is, the self-determined
coordination of the social collective—have a future. The problem, however, can also be stated exactly the other way around: only those forms
of socialization open to negotiation among different collective modes
of life, each with their own ways of socially constructing society's relationships to nature, through a negotiation aimed at coordinating among
these different modes and perhaps generalizing from some of them—in
short, only self-determining forms of socialization—will be in a position
to prevent a usurpation of temporal horizons stretching far into the
future and the resulting ecological crisis.

In what follows I want to discuss two approaches found in current
critical social theory to the problem of constructing an ecologically
sound constitutional theory. The two approaches, by Thomas Blanke
and by Ulrich Preuss, respectively, attempt to integrate the perspectives of both ecology and democracy, and to contribute theoretically
to the development of democratic institutions that would have a depth
of intervention capability sufficient to enable them to steer economic,
technical, and ecological developments, while taking into account the
long-term consequences of social development as a whole, including
state intervention. According to these approaches, furthermore, state
intervention should be subject to collective decision making.

The goal, then, of these proposals is to allow constitutional theory, with its protected freedoms and procedural norms, which together guarantee a democratic coordination of different interest positions by stabilizing the expectation and decision horizons of social
actors, to take into account changes in the entire social system to
which it applies. New problems, new social relations, and new models
of action are not easily dealt with by systems of norms that have been
handed down over the years. The two proposals argue, therefore,
for a new stage of development of the bourgeois state, one in which
the negative consequences of welfare state intervention can be reflexively integrated in both the decision and implementation processes
of government, and where the state is committed to ecological values.

I will first argue that neither proposal discussed here can solve the
problem (given a capitalist society) of integrating state intervention,
democracy, and ecology. Then I will suggest some ways in which a
Marxist-oriented theory of democracy can contribute to a resolution
of the crisis in the way society's relationships to nature are currently
being socially constructed.

However, I want to make clear from the outset that I share
Blanke's and Preuss's assumption that such contributions to demo-

cratic theory make sense because "time is pressing," and that therefore it makes little sense to pursue a model of revolution that presumes the resolution of all social problems—in this case, ecological problems—will follow from the abolition of capitalist relations of production and their replacement with socialist relations of production. In contrast, a strategy aimed at resolving ecological problems themselves would be much more likely to contribute to a democratic transformation of public discussion and collective decision-making processes now subjected to a particular socialization by capitalist relations of production.

3. Blanke's Procedural Approach to Incorporating Ecology in Constitutional Law

One possible way of integrating ecological points of view into democratic procedures and governmental decision-making and implementation processes consists of extending constitutional principles and the rule of law to nature. Thomas Blanke has attempted to work out this option in a preliminary study concerning the way to construct an ecologically sound constitutional theory.

Rejecting a critique inspired by Herbert Marcuse, who sees the realm of law as sanctioning nature only in terms of an already existing socialization of nature, one in which nature is socialized as subordinated to the existing mechanisms for valorizing capital, Blanke emphasizes the immanent relationship between nature and society: given the highly developed state of society's domination of nature, nature has for a long time been a "moment" of society. If the negative consequences of society's present ways of socially constructing its relationships to nature are to be overcome, then nature, which up till now has been an area excluded as an object of legal regulation, must be consciously included in such regulation. In this way, Blanke wants to relocate the problem. It has now become a question, as he sees things, of how we can make the legal system dynamic enough to be up to facing new areas of social problems—in this case, ecological ones.

The process of social development, according to Blanke, is not defined by systemically determined, unchangeable laws. Rather, it is mediated by the actions of social actors committed to the principles of a universalistic morality to which they can refer in cases of conflict. A social order founded on constitutional principles of freedom and equality, and requiring social actors to rationally justify their actions, cannot be conceived "as a concrete order, one with a particular set of correct contents"; rather, it must be understood "as simply a process of procedures, open as far as results are concerned, and involving the equal participation of all concerned."[5] Existing rationally and legally founded constitutional principles guarantee in themselves a sufficient reservoir of rationality upon which democratic decision-making procedures can draw when needed. They offer, that is, the guarantee of normative criteria, which can be drawn upon to check the writing into law of specific social relationships that would threaten the freedom and autonomy of individuals. In this way, such normative criteria can preclude, in principle at least, the establishment of legal procedures that are self-binding in terms of contents and substantive decisions, and guarantee a high degree of "reversibility"—that is, an openness to the future that gives newly formed interest groups enough time to fashion themselves into a majority, and with the power to reverse previous decisions. (This is obviously significant, for example, with respect to the construction of nuclear power plants and the destruction of ecosystems.)

At the same time, calling upon the counterfactual, hypothetical norms of a rational morality does not represent a threat to the legal order because these norms are already incorporated in the constitutional order, for they act as a legitimation of the legal system. By building a legitimation of the legal order into this very order, a new dynamic at a higher level is created. For the established system of legality is constantly being challenged by the claims of individuals calling upon the system's own principles of freedom and autonomy, and thus prodded into further development—or, if the system fails to develop, calling serious crises down upon itself.

Blanke argues, moreover, that the internal dynamic of modern, bourgeois legal systems, based as they are on rational, legal principles, permits one to expect that the systems will be capable of themselves generating the means for overcoming forms of domination. This is the case because the legal system is constantly developing itself at a higher level in order to prevent the uncoupling of legality and legitimation.[6]

Blanke sees a danger existing for this historical dynamic in the particular conception of legitimacy held by the FRG's Federal Constitutional Court (a criticism that has been made for years by others). The court's bundling together of fundamental constitutional principles, however progressive these may be, into something called "a liberal democratic constitutional order," is a move that "submits the self-reflective determination of society's reproduction to a canonical set of binding norms," and which reveals "the power to define the contents and limits of democratic legitimacy to be a state monopoly."[7] This tends to invert, Blanke argues, the democratic process of constituting society. It leads not only to a hindering of the unfolding of the principle of democratic self-government, but to its very disappearance behind a given and unchallengeable body of constitutional law. But the principle of democratic self-government can only be preserved in the face of a hasty and politically motivated legal codification of interests if procedures remain open and reversible.

I would now like to take a critical look at Blanke's ideas. Blanke is able to show in his discussion of a rational, legally founded, and procedurally structured constitution that the bourgeois principle of law cannot be viewed simply as one more means of oppressing nature and expanding its socialization. However, it is not at all clear in what sense such a constitution can be understood as specifically ecological. For a constitution conceived according to rational-legal principles can at most raise a "normative barrier against industrial 'autism' vis-à-vis nature."[8] But since legally acceptable principles do not on their own determine substantive outcomes, barriers will be raised only when, and to the extent that, citizens raise their voices for the sake of ecological "interests." Blanke's position allows neither for a substantive orientation of the constitution in terms of ecology, nor for an automatic taking into account of ecological points of view during formal legal procedures. In fact, "nature" does not need to be considered at all, since, according to Blanke, it is already socialized. It can be concluded, therefore, that a proceduralistic constitutional system based on principles of rationality is not capable of dealing with the contemporary ecological crisis.

Blanke regards the tension between legitimacy and legality found in existing constitutional systems to be so dynamic as to enable these systems to react appropriately to ecological problems—provided specific values are not written permanently into a constitution. However, one can have doubts about the dynamic dimension of constitutional systems whose learning processes are faced with problems having a much longer temporal perspective. Blanke assumes that the law is constantly in flux, driven by its continuous attempt to make good on the claims of its internal principles to an ever higher evolutionary level of social learning. This way of constructing the argument relies on historical–philosophical, finalistic, and teleological assumptions to such an extent that the development of the system of law seems to occur exclusively by means of the inner dynamic of the postulates of freedom and equality intrinsic to the bourgeois conception of rational law.

However, this way of arguing constrains the possibilities of social learning, since the latter are restricted to the single dimension of realizing the principles of freedom and equality. It is undeniable that modern bourgeois society is characterized by an opposition between normative claims and a reality that does not live up to these claims. But this alone does not establish a truth that the resulting dynamic will in fact lead to an ecological and democratic "improvement" of capitalist society by means of realizing these norms. For the relationship between norm and reality can itself be conceived as a specific, social relation of capitalist society.

The dynamic that Blanke describes is therefore to a certain extent an "empty" one—but one that all the same raises some very fundamental political problems. Debates within democratic theory have shown that it is precisely the attempt to realize freedom and equality, as these ideals have been understood in bourgeois society since the French Revolution, that turns into totalitarianism. For this reason, Marx, from the time of his critique of the Jacobinists, distanced himself from viewing socialism as the realization of the norms of equality and freedom.[9]

From this critical perspective, a multiplicity of problems come into view that can be subsumed under the heading "ecological crisis," problems that arise precisely out of the dynamic of progress, and the logic of evolution, which is reproduced anew, at every level of capitalist development, within the tension between norms and reality. Critical Theory's critique of progress, despite its commitment to early bourgeois ideals, always strongly emphasized the idea that there must be a break with past conceptions of "progress," in other words, that there must be a progress in our conception of "progress" itself.[10] Bourgeois law, however, can only operate within the tension between rational–legal norms and positive law. And since bourgeois law is, according to Blanke, dependent on socially mediated explications of, and concrete calls upon, these norms, it does not have control over the parameters of its own dynamic. A theory of law that does not take this critique of the legal codification of the tension between norms and reality into consideration binds, from the very beginning, the form of social learning in capitalist societies to the realization of bourgeois standards of legitimacy.

In passing I would like to point out that Blanke's position does give a theoretical articulation to the conception of democracy widely espoused by West German Greens. The Greens criticize the dominant form of parliamentary–representative praxis because it blocks the participation of all citizens in an open and public process of discussion and decision making (*Willensbildung*). Their goal is to open up, or create anew, parliamentary and representative channels by which all opinions and interests can be brought into the discussion with as little distortion as possible. Only through this kind of opening up of the political system, it is thought, can a sufficient measure of flexibility be created which can be used to respond effectively to new, above all ecological, problems.

The Greens' theory of democracy assumes—rather questionably—that a higher degree of openness in the public process of discussion and collective decision making will lead to a more thorough consideration of ecological interests, because people's immediate "life interests"—that is, both their survival interests and their interests arising from everyday life—can be brought to bear on the process. The assumptions at the base of this political theory are doubtful in several respects. First, the theory assumes that the closing of the gap between bourgeois constitutional norms and the social reality of the German Federal Republic—a dynamic I have already discussed from the perspective of the critique of ideology—has led already to such a high degree of openness with respect to government institutions that a reorganization of democratic procedures and institutions is not necessary, and decisions concerning long-term ecological problems are possible. This simply ignores the problem of implementation and state management.

Second, a classical dilemma of the theory of democracy is ignored, and thereby reproduced, when it is assumed that a democratic unfolding of interests will more-or-less automatically lead to a commitment to an "ecological common good." In other words, the assumption here is that an "undistorted" majority decision is also a "rational" decision in some ecological terms.

Third, the Greens' theory of democracy ignores the fact that a reliance on life interests can also have antidemocratic consequences. For it means a naturalization of politics to the extent that these interests are not themselves being made the object of a democratic process of discussion and negotiation.

4. Preuss's Proposal for an Ecologically Self-Reflexive Constitution

Blanke's approach involves a theoretical strategy that regards the ecological crisis as manageable within the framework of a democratic constitution outfitted with an expanded procedural rationality with respect to the rule of law (*Rechtsstaat*). Ulrich Preuss follows the exact opposite strategy in his approach, which I would now like to discuss. He sees it as imperative that phenomena characterizing ecological crises enter directly into the attempts at resolving these crises, and that these phenomena in their "peculiarity also [determine] the character of the solution to problems."[11]

.   In Preuss's reconstruction of the problem, the function of a constitution embodying the rule of law has been, until recently, to mediate among conflicting social interests by means of procedures and institutions that would transform social interests into political rule and rationalize them. Now, however, one type of "left" constitutional theory is no longer sustainable, namely, the theory that geared itself to the growth-oriented class compromise of the past decades and that strove to maintain or expand the capability of the working-class movement to articulate and push through its interests without regard for the externalization of all the negative consequences of growth, and for their accumulation in the form of crisis tendencies. If "democratic emancipation and self-determination is ever to develop itself once again as a project capable of effecting change,"[12] Preuss argues, it is necessary to win time and to hold the future open for revisions of previous decisions and for new democratic options. Preuss suspects, as does Blanke, that this will only be achieved if the capacity of the constitution for institutionalized reflexive learning can be increased by means of a new institutional design, so that the conditions governing the processes "in which society constantly revolutionizes its own basic structures" can be taken into account. The constitution would become reflexive in that, flexibly reacting to new problems, it would itself, in accordance with its own principles, constantly be changing. In Preuss's words, "The constitution itself would be continually undergoing change so that one could say in sum that constitutional theory is a theory of controlled constitutional change."[13] Constitutions, then, would not only be open to new problems but, with a more far-reaching temporal perspective aimed at reflexive learning, they would also be more able to assume the function of steering global social development, a role that the law in the course of the history of bourgeois society has perhaps always tried to assume but so far without any success.

Only such a "reflexive constitution," one structured by procedures that would subject its own procedural process to democratic decision making, would guarantee a sufficiently high level of learning and the required openness to the social problems of the future. In this, however, one of the decisive desiderata of modern bourgeois democracy is met, that of the reversibility of decisions. The difficulty faced is that, even if no catastrophe occurs, the effects of the introduction of new technologies, the exploitation of natural resources, and the destruction of the environment are all substantially irreversible, and hence are incompatible with the "reversibility" principle of ratio-

nal and democratic decision-making procedures. A possibly historically transitory mode of production, and a specific configuration of
constitutionally embodied social decision-making mechanisms, determines and limits the future of many generations—in other words,
renders reversibility practically impossible. However, Preuss argues,
with the model of a constitution-become-reflexive, such risks would
no longer be unloaded onto future generations. In fact, these generations would be institutionally included in today's democratic decision-
making processes as silent participants.

Up to this point, Preuss's argument is entirely procedural in
nature. He gives no reason why the rationality and capacity for social
learning of a more highly developed constitution should result from
reference to an ecological parameter. Preuss himself makes this clear
in his discussion of the concept of an ecological democracy. Here he
argues, rather trivially, that human beings as social beings place use
claims on the natural economy, with the result that they are threatened
as natural beings. The capacity to learn in this case refers only to the
reflective consideration of justifiable human interests. This would
lead, as in Blanke's argument, to the avoidance of ecologically unsound political decisions by means of an active electorate and a high
level of voter participation. However, this view assumes precisely
what cannot be assumed in a discussion of constitutional theory,
namely, that social groups and individuals would want to constantly
involve themselves with ecological problems and their solutions. At
the same time, legal codification is, above all, supposed to render this
kind of continuous engagement superfluous, and to relieve social
actors of their burden, by assuming a positive steering function.

Only with two further arguments, which make the concept of
reflexivity with respect to ecology more precise, does Preuss give
some indication of the positive characteristics of a specifically ecological democratic constitution. First, in his opinion, the relationship of
human beings to nature can be ecological only to the extent that
"human beings see themselves as part of the natural economy and
organize their relations of exchange and interaction with the other
elements of this economy according to a principle of the reproduction
of the whole system." Beyond a mere strategy of avoiding damaging
interventions in natural processes, the concept of ecological democracy describes the problem of a form of human socialization that has
become conscious of the fact that it is a part of integrated ecological cycles.

Preuss himself does not develop this argument further; and, so
far as he has elaborated it to date, the objection can be made that
instead of considering the degree to which nature has been, and is

being, socialized, Preuss's argument dissolves society into nature. The assimilation of ecology with the modern constitutional state based on the rule of law (*Rechtsstaat*) would negate the properly historical character, and hence negotiability within society, of legal norms— with the danger arising that every violation of the laws of the state would assume the fateful character of a violation of the laws of nature.

The view Preuss seems to take is that social life should be structured in line with ecological organizational principles. This begs the questions of what principles, and what understanding of "ecology"? According to Preuss, social relations are to be structured so as to benefit the unfolding of human civilization. By "beneficial" is meant, drawing on an analogy to ecosystems, a successful combination of the preservation of the achievements of civilization together with the maintenance of the "unbounded potential for change" that results from the flexibility of human beings. Also analogous to ecosystems is the idea that a high degree of cultural diversity is a sign of a high degree of flexibility and a large capacity for change. Ecological democracy will, Preuss asserts, contribute significantly to the maintenance of this flexibility, and it is this flexibility that allows human civilization "to steer its self-transformation."

Preuss, then, wants to go beyond the view that a constitution functions only as a means for balancing conflicting interests and of coordinating the proposals for the future put forth by different power blocks. Social changes resulting from the competition among power centers cannot be viewed, according to him, as learning processes. The model of checks and balances, of power and counterpower, generates an incapacity to learn, and does not guarantee sufficient openness and flexibility in the sense of "a potentiality for transformation." Ecologically structured democratic institutions have to get used to dealing with uncertainty, "that is, to an unknown quantity and quality of bits of new information that we have to process." And they have "not so much to react to or against changes as to—reflexively—acquire the capacity for self-transformation as one element of social transformation."[14]

It becomes clear that, for Preuss, only an ecologically oriented constitution attuned to uncertainty, one that is situated and operating "underneath" the different interests groups, can mobilize a sufficient degree of learning capability within political and social institutions, and can therefore organize "the permanent transition from one unstable situation to another." Given the present ecological crisis, there are no longer any external strategies of avoidance open to politics—including centralized political steering mechanisms—that could create stability by establishing a basic framework for governing social action while ignoring the diversity of life worlds (*Lebenswelten*). Rather, the conditions of action must themselves be organized in the respective social contexts of action. A total social rationality no longer exists: "What is characteristic of an ecological social model, therefore, is the self-coordination of heterogeneous life worlds rather than their subjugation to a homogenous, unified will."[15]

Preuss, with these proposals for an ecological democracy, has ventured further than Blanke; and in doing so, he has left somewhat more certain ground of established legalistic-proceduralistic constitutional theory, in the effort to develop new extensions and elements of this theory.

While raising certain objections to this project, I want to make clear that I share the assumption made by Preuss, and by Blanke as well, that every form of externalization of the present form of social life onto external nature, the state, and the future is obsolete. Capitalist societies have been once and for all confronted with the fact of their immanence (*Immanenz*). This counts for nature, which a long time ago (as noted already by Marx and Engels in their *German Ideology*) became an integral part of this social immanence (*gesellschaftliche Immanenz*). And it counts also for the state, merely a "mortal god," which can no longer be taken as an extrasocietal and central authority (*Instanz*) that steers society. That said, I would now like to raise five objections to Preuss's argument, with the aim of showing that he has not achieved the goal he has set himself, namely, to integrate the three aspects of state management, democracy, and ecology.

First, Preuss's ecological perspective misleads him into taking a harmonistic view of nature. For, while relying on an analogy with the variety of species, he assumes that a "social ecosystem" containing a multiplicity of life worlds would be better able to deal with uncertainty than a system for negotiating proposals for the future that are based on interests. But one should remember that the "natural" ecological balance is a precarious one, resulting as it does from a multiplicity of intertwined food chains. Nor should one forget that the potential for change and flexible adjustment characteristic of the variety of species implies also that individual species disappear from the ecological cycle. And, finally, it should not be forgotten that stable ecosystems, inasmuch that they may be described in terms of a functional equilibrium, inscribe each species to definite functions within the whole. Adopting ecological metaphors for social theory is, for this reason, rather problematic.

Second, the theoretical supposition that nature solves the problem of uncertainty as such relies on holistic and quasi-theological premises for its plausibility, premises that begin with some notion of creation and nature as such, and not with individual species. Carried over into social theory, this leads to the view that (eco)systemic interests take priority over particular forms of life, even though it is assumed at the same time that the continued existence of society depends on the form and generalization of particular modes of life and models of reproduction.

Preuss's argument by analogy is therefore circular. To the extent that nature is holistically conceived and presented as a model for social relations, it is projected as being external to society, and then enlisted as an orientation for social reproduction. In this manner, society defines its relation to itself by means of nature as an external medium. And to the extent that nature becomes a norm for societal orientation, it also becomes a neutral and neutralizing area of social contradictions and conflicts.

Third, why should one adopt the perspective of the total system of a society, as suggested by the concepts of uncertainty and learning? Learning to what end? Is it appropriate to describe our social relations as uncertain? One could just as well argue the opposite: the political developments that led to the promotion of nuclear energy production and, more recently, to reproductive and biotechnologies, are rather well known. They can be explained well enough in terms of an analysis of capitalist relations of production. To prevent or contest such developments does not require a focus on uncertainty. A focus on uncertainty would only be required in the limited sense of acknowledging the probability of catastrophes in already existing nuclear power plants, or of having experimentally created viruses run wild, or of the destruction of the rain forests, and so on. Such considerations would lead, if ecological prudence were the criterion, to the oft-called-for moratorium on, or complete stop to, such investments. So the real problem here has to do with the long-term effects of such projects, about which there is little uncertainty as such.

Fourth, what also seems problematic is the argument for an orientation toward the future. Here the problem seems, to me, not only to involve a question of moral philosophy—namely, why should an orientation toward future generations be made a binding duty of contemporary action—but also the construction of the opposing poles of an openness to the future versus a contemporary autism, that is, "system openness" versus "system blockage." Speaking against the construction of such oppositions is the fact that every generation imposes the consequences of its premises on following generations. Since such a "fixing of the future" always and necessarily occurs—because the passing of generations is not a zero-sum game—standards of evaluation can only be set according to the criteria of a given historical period. In any case, there is no one homogeneous future shared by a homogeneous, simultaneously present society; rather, classes and social groups have many different futures that play themselves out one against the other. A moral appeal will not have much effect against corporations that are working to realize a specific "fixing" and unfolding of a future answering to their particular needs. The future, nature, and the human species are not, here, able to play the role they are asked to play, of substantial criteria of democracy—no more than (as Preuss rightly points out) could the contrived assumption, as in traditional Marxism of a set of homogeneous working-class interests.

However, despite Preuss's problematic substantialization of nature, the species, and the future, it seems nonetheless plausible to me that the future as well as nature have become, in an irreversible way, the object of political decision making, and that they no longer can be ignored in a democratic public process of discussion and collective decision making. But this is not a problem of the future; rather, it concerns the present. Moreover, it is not a matter of a zero-sum game, but instead an attempt to transform a perceived negative-sum game into a positive-sum game.

Fifth, the system-theoretic implications of Preuss's argument also become evident when he suggests that the political system, which until now has sought to coordinate the interests and actions of power groups, be reorganized so as to be able to permanently recalibrate its state of balance. There seems to be some self-contradiction here. The problem of regulation is supposed to be solved by means of a self-coordination on the part of life worlds. But it does not seem plausible that someone can assume, simultaneously, the standpoint of a specific life world and that of the coordination of all life worlds—even more so, when this latter standpoint is no longer supposed to exist (being subsumed within nature). Preuss cannot avoid this contradiction, however, because he does not want to give up the claim that constitutional theory can solve the problem of the self-coordination of diverse life worlds and decentralized public spheres.

5. Toward a Concept of Society's Socially Constructed Contradictory Relationships to Nature Involving a Play of Domination and Emancipation

The theoretical proposals for a further development of the bourgeois state that seek to include a reflexive rationality that encompasses ecological parameters prove to be problematic and self-contradictory.

Again and again, one aspect or another of the "magic triangle" of democracy, ecology, and state management slips from one's fingers. For as soon as nature is made over into a relevant parameter of state action, the democratic element comes up short. On the other hand, if one puts the accent on democracy, it is no longer clear how state management is supposed to be possible.

Despite these problems, it is still reasonable to attempt to develop proposals within the theory of democracy for rationalizing state action—even if they have only a restricted range of validity and their chances for successful implementation are rather slim. For it can be assumed that the bourgeois state is a condensation of unstable balances of power arising among social classes and movements—or, put differently, a form of movement of social contradictions. Therefore, it is both possible and advisable for subalternate social collectives, the labor movement, and social protest movements to each develop, with a view to their futures, their own standards of rationality and forms of social coordination, making these, in turn, a factor in the precarious balance of compromises. In this way these social groups and movements can codetermine the form of state domination arising from the present crisis.[16]

In passing, I would like to emphasize that, in speaking of rationality criteria here, I do not mean a total social rationality, with the help of which one could criticize the bourgeoisie for enthroning its particular interests as the common good. Such a critique of ideology alludes only to the significance of the unintended side effects of a capitalist-determined appropriation of nature; it ignores for the most part, however, the problem of cause—namely, the appropriation of surplus value in the process of production, and the dynamic of accumulation, with its "naturelike" (*naturwüchsig*) process of the unfolding of the social division of labor.

This line of critique of ideology assumes, in the perspective of a sociology of domination, that the diversity of social interests is the result of a divide-and-conquer strategy on the part of the ruling classes, which seek to prevent the unfolding of a homogeneous general will on the part of the dominated. What is being overlooked here is that standards of rationality are themselves class-, group-, and gender-specific. In contrast, the "critique of rationality" discussion stemming from the work of Horkheimer and Adorno, as well as that of Foucault, has made the political significance of a concept of rationality clear; and has also drawn attention to the fact that an essential step toward democracy and an overcoming of capitalist relations of production consists precisely in the construction of novel forms of the generalization of social interests. Above all, this project must aim at opening up the current form of the social division of labor to collective social decision making. This, after all, is the area in which the largest proportion of individual and collective interests arise and are satisfied, as well as being the place where society's social construction of its relationships to nature assumes its technical form.

Despite my agreement with Preuss and Blanke on several points, I would raise the fundamental objection that both authors fail to bring together the concept of nature, as embedded in socially constructed relationships to society, and relations of production. In other words, they leave out of view the significance of capitalist relations of production for the concrete form taken by society's socially constructed relationships to nature. Blanke rejects the kind of critique of capitalism, à la Marcuse, that relies on using a critical concept of an undistorted nature. Preuss locates the problem of ecology beneath the relations of production and class relations, at a subpolitical level. He tries, that is, to constitute at a new level a total social rationality, one that can circumvent the negative consequences of economic growth based on social compromise.

In the end, both authors come to the conclusion that the further development of the bourgeois state could take the form of a controlled constitutional development as a learning process. In this development, constitutional law and the jurists who conceive and interpret it are assigned the role of initiators. They are—here translating Blanke's and Preuss's arguments into a scientific–juridical ethic—responsible for guaranteeing the continuity and stability of the legal order and, at the same time, for ensuring through their initiatives that a dynamic, self-transforming constitutional praxis can initiate, steer, and determine future developments. Thus, it turns out that the historical development of capitalist societies depends essentially on lawyers and jurists. And, moreover, emphasis is clearly given to law, and not democracy, as the motor of change in capitalist societies.

These overgeneralizations of constitutional theory as the motor of development for the whole of society lead, de facto, to a narrowing of the problem of an "ecological democracy" to the realm of the constitutional state (*Rechtsstaat*). I would like to insist, by contrast, that society's social construction of its relationships to nature cannot be separated from its relations of production and their inherent social conflicts. Following a strand of discussion originating in Western Marxism, I think it makes more sense to assume that each of the different and antagonistic social groups and classes, depending on their position in the social process of production, has their own socially constructed relationship to nature. This argument, in particular as found in Gramsci's "Quaderni del carcere" and Horkheimer's and

Adorno's "Dialectic of Enlightenment," builds on Hegel's view that the form a relationship to nature takes is determined by the social division of labor in which it occurs—paradigmatically, by the master–slave relationship. Only for the master is nature a pure object, one that he can appropriate immediately because the slave makes it, in an always-already worked-up form, available to him. The self-consciousness of the slave, on the other hand, is mediated through work and the process of forming natural objects.[17]

Society's metabolic exchange with nature takes place within society itself. While it was possible for a long tradition within political philosophy to believe it could locate the values of good government in an adaptation to nature, since Hegel and Marx—at the latest—it must be assumed that social and political relations are not, and cannot be, prescribed by nature. Nature as such is not a unified whole existing in some original balance; and the natural history of human societies is contingent and reversible.

But even assuming such a unified and balanced nature, there is still the need to interpret this theoretically. Thus, social conflict would not be ended by such an assumption; rather, it would be carried over into an endless conflict over the best way of adapting to a (theoretically) imagined natural balance within nature.

In contrast to this way of seeing things, and with a view to the abolition of domination, it seems more democratic to begin with the insight that society's socially constructed relationships to nature are bound up with relations of domination. Gramsci, as well as Horkheimer and Adorno, worked out in different ways the idea that the social division of labor, and the domination and exploitation bound up with this, constitutes for different classes a different form of unity with nature, a different "block." Horkheimer and Adorno, still trapped within Lukács's subject–object philosophy, could only find in these "blocks" the total reification of alienated labor, which in the end leads to the complete socialization and subjection of primary nature (*der ersten Natur*). Gramsci, however, emphasized above all that the socially constructed relationships to nature assumed by different classes always stand in a relationship to hegemony and compromise. Thus alternative forms of socially constructing society's relationships to nature embodied in lower classes, dominated peoples, and social collectivities have not (yet) been able to unfold, and exist only in embryo.

"Nature," in short, is the historically contingent result of the generalization of class-, group-, and gender-specific ways of socially constructing society's relationships to nature. This generalization takes place in, and through, the collisions that occur at all levels of society—the economic, the political, the scientific, and the technical—and also in, and through, the implicit and explicit compromises found at all levels if there is to be any continuity to the capitalist process of accumulation at all.

One can conclude, then, that the present crisislike situation is the result of, among other things, the fact that the Fordist–Keynesian class compromise that existed over the past decades has been broken up by the emergence of new forms of life and new social actors in many new socially constructed relationships to nature. This is the case not only for the emergence of the new social movements in the shadow of the great corporatistic blocks. The entire set of standards of rationality and objectivity constitutive of the Fordist phase of accumulation has also been put in question by the critique of universalistic moral principles developed by the women's movement.[18]

Since "nature" can no longer be empirically and naturalistically determined, and proves to be instead a theoretical construction, the forms of scientifically constructed objectivity become themselves the objects of genealogical analyses (as in Foucault and Fox-Keller, among others) and are drawn in this way into intense conflicts. Conflicts over the hegemony of specific scientific constructions of objectivity can no longer be resolved with the instruments of scientific argumentation; rather, these must be resolved with the help of official–symbolic force. This became particularly clear in the FRG after the reactor accident in Chernobyl, when after a conflict arose concerning the tolerance levels for radiated food a new Ministry for the Environment was created, which then tried to claim an official monopoly in terms of defining tolerance levels.[19]

According to the position presented here, then, there is no single way of constructing society's social relationships to nature; rather, there are many such ways, which stand, even in the phase of the Fordist model of growth, in hegemonic relations to one another, constituting a balance of compromises. The level of political compromises among power groups, and the level of varied modes of life cannot legitimately be separated form one another—and this is what Preuss tries to do, with his assumption that the heterogeneity of life worlds (*Lebenswelten*) as such is already able to produce an ecological balance attuned to nature. Rather, the form that the compromise among society's actors assumes will define as well the dominant way in which society socially constructs it's relationships to nature; the way nature is appropriated; the form of the labor process and its organization within the social division of labor; and the form of political coordination and regulation.

In other words, if one wants to integrate ecological parameters as integral elements into a theory of democracy, then one must push forward to a democratic coordination of society's socially constructed relationships to nature. But "nature" is a social category—that is, society's socially constructed relationships to nature are determined by the labor process, and by the process of appropriation, as these occur within forms of social cooperation. For this reason, democratic decision-making mechanisms must necessarily have jurisdiction over the social division of labor, as well as over the process of differentiation of autonomous areas of activity as determined by the logic of accumulation, thereby subjecting these to a social decision-making process.

One consequence of this way of approaching the problem of "ecological democracy" is that the boundary lines between heterogeneous modes of life are put in question. For social and ecological problems arise, to a great extent, from the naturelike (*naturwüchsigen*) character of the dynamic of social cooperation, which differentiates out various social activities in a manner that has a natural lawlike character, making them blind to one another. In the face of the horror felt by proponents of systems theory, and theories of modernity, toward a questioning of the logic of differentiation, it must be insisted that differentiation need not remain the final and "natural" law of development for modern societies, one that is beyond the control of social actors.

However, from the perspectives of a theory of democracy, the goal cannot be the manufacture of a general will. As an alternative, and as an escape from this trap of modernity, one can turn to a "culture of consensus," which can be achieved by means of a new logic of articulation.[20] Accordingly, an overcoming (*Aufhebung*) of the division of labor does not mean—and here I follow an often misunderstood point made by Marx—a regression from the high level of social cooperation already achieved in capitalist society. Rather, it means just the opposite: social cooperation would be increased, in that the form of social cooperation itself would be made the reflexive object of a self-determining coordination. A rearticulation of cooperative fields of praxis represents an increase in rationality and complexity to the extent that the form of the social division of labor would become reflexive in the direction of both social differentiation and dedifferentiation, and would come under the control of social actors.

Such a strategy of reflexive social differentiation and dedifferentiation could be used, for example, to systematically reduce the accumulation and concentration of decision-making competence and necessity at the top of government, or corporate, hierarchies. Problems could be dealt with instead where they arise; effects and side effects could be quickly and flexibly drawn into the decision-making process; and, depending on the problem, the form of social relations and their respective socially constructed relationships to nature could be flexibly reconstituted. And precisely this greater degree of development of the social division of labor and cooperation would prevent a damning of individuals to lifelong forms of activity or social roles (e.g., class or gender roles). Put positively, it would be possible for individuals to freely associate, and through this association let their interests unfold, since they would have direct access to the social division of labor; to the ways in which society's relationships to nature were being socially constructed; and to their relations to one another.[21]

Notes

1. Cf. James O'Connor, "Capitalism, Nature, Socialism: A Theoretical Introduction," *CNS*, *1(1)*, no. 1, Fall 1988.
2. Niklas Luhmann, "Die Zukunft der Demokratie," *Soziologische Aufklärung*, 4 (Opladen, Germany: Westdeutscher Verlag, 1987), p. 127 and infra.
3. Rainer Wolf, "Zur Antiquiertheit des Rechts in der Risikogesellschaft," *Leviathan*, no. 3, 1987; see also Thomas Blanke, "Recht System und Moral—Vorüberlegungen-zu einer ökologischen Verfassungstheorie," in H.-E. Böttcher, ed., *Recht Justiz Kritik* (Baden-Baden: Nomos Verlag, 1986).
4. Ulrich K. Preuss, "Die Zukunft: Müllhalde der Gegenwart?," in Claus Offe and Bernd Guggenberger, eds., *An den Grenzen der Mehrheitsdemokratie* (Opladen: Westdeutcher Verlag, 1984).
5. Blanke, "Recht System und Moral," op. cit., p. 402.
6. Ibid., p. 414 and infra.
7. Ibid., p. 417.
8. Ibid., p. 402.
9. Karl Marx, *Grundrisse zur Kritik der Politischen Ökonomie* (Berlin: Dietz Verlag, 1953), p. 160. Also see A. Maihofer, "Ansätze zur Kritik des moralischen Universalismus," *Feministische Studien*, no. 1, November 1988; and Alex Demirović, "Marx und die Aporien der Demokratietheorie," *Das Argument*, 172, January–February 1991.
10. Walter Benjamin, "Der Begriff der Geschichte," *Ges. Schriften*, vol. 1.2 (Frankfurt: Suhrkamp Verlag, 1974); Max Horkheimer, "Autoritärer Staat," *Ges. Schriften*, vol. 5 (Frankfurt: S. Fischer Verlag, 1987); Max Horkheimer and Theodor W. Adorno, "Dialektik der Aufklärung," in M. Horkheimer, *Ges. Schriften*, vol. 5 (Frankfurt: S. Fischer Verlag, 1987).
11. Ulrich K. Preuss, "Die Zukunft der Demokratie," *Komitee für Grundrechte und Demokratie*, Jahrbuch, 1987, p. 158.
12. Ulrich K. Preuss, "Aktuelle Probleme einer linken Verfassungstheorie," *Prokla*, *61*, 1985, p. 77.

13. Ibid., p. 78.

14. Claus Offe, "Die Staatstheorie auf der Suche nach ihrem Gegenstand. Beobachtungen und Diskussionsstand," in Thomas Ellwein, Jens Hesse, Renate Mayntz, Fritz Scharpf, eds., *Jahrbuch zur Staats und Verwaltungswissenschaft*, vol. 1 (Baden-Baden: Nomos Verlag, 1987), p. 13.

15. Preuss, *Komitee für Grundrechte*, op. cit., p. 14.

16. Alain Lipietz, *Trois crises: Metamorphoses du capitalisme et mouvement ouvrier* (Paris: Cepremap, 1985); Alex Demirović, "Bürgerliche Demokratie—Ein historischer Kompromiss?," *Archiv für Rechts- und Sozialphilosophie*, no. 4, 1987.

17. Gottfried F. W. Hegel, *Phänomenologie des Geistes* (Frankfurt: Suhrkamp Verlag, 1970), p. 150 and infra.

18. Cf. Maihofer, "Ansätz zur Kritik," op. cit.

19. Cf. Ulrich Beck, *Risikogesellschaft. Auf dem Weg in eine andere Moderne* (Frankfurt: Suhrkamp Verlag, 1986); and Wolf, "Zur Antiquiertheit des Rechts," op. cit.

20. Cf. Ernesto Laclau and Chantal Mouffe, *Hegemonie und radikale Demokratie. Zur Dekonstruktion des Marxismus* (Vienna: Passagen, 1991); English version: *Hegemony and Socialist Strategy* (London: Verso, 1985); see also Alain Lipietz, *Les Conditions de la construction d'un mouvement alternatif en France* (Paris: Cepremap, 1986).

21. Cf. John Burnheim, *Über Demokratie. Alternativen zum Parlamentarismus* (Berlin: Wagenbach Verlag, 1987), p. 181; English original: *Is Democracy Possible?* (Cambridge, England: Polity Press, 1985).

Index